438429

D1307785

Conservation of *Natural* RESOURCES

ESS 108 | Kay R. S. Williams

Kendall Hunt
publishing company

Kendall Hunt
publishing company

www.kendallhunt.com
Send all inquiries to:
4050 Westmark Drive
Dubuque, IA 52004-1840

Contents

Chapter 1

Introduction to Conservation of Natural Resources

BASIC CONCEPTS

I.   ***Types* of Natural Resources**—any substances and energy sources humans need to survive (see Figure 1.1)

     A.   *Inexhaustible*—perpetually renewed; for example: solar, geothermal, and wind energy

     B.   *Renewable*—replenished over short time; for example: fertile soil, clean air, and wildlife

     C.   *Nonrenewable*—finite supply; for example: fossil fuels, minerals, wilderness

Tupungato/Shutterstock.com

II.   **Environmental Science Themes**

     A.   Human population growth—the major environmental problem

     B.   **Sustainability** (Humans living in a way to maintain Earth's systems and natural resources)—the environmental goal

     C.   Global perspective

     D.   Interdisciplinary (see Figure 1.5)

     E.   People and nature

     F.   Science and values

     G.   Science provides solutions; which chosen partly value judgments, including **ethics** (moral principles or values)

III. *Time* **(see App. A)**

    A. *Geologic*—Earth's 4.5 billion year history (see App. 1)

    B. *Biologic*—time periods less than10 thousand years

        1. Limited by periodic, major climate changes

        2. Oldest known living organism—the bristlecone pine tree of CA, ~5000 years

        3. Organisms slowly adjust to changing environmental conditions

        4. Humans accelerate rate of change to environment

        5. Organism don't have time to evolve adaptations to adjust to changes

        6. Can't survive if unable to move to area of favorable climate

IV. **Major viewpoints toward natural resource use**

        1. *Anthropocentrism/Exploitation*—use all humans want (e.g., Clear cutting the forest)

        2. *Ecocentrism/Conservation*—use wisely; leave some for future generations (e.g., Select cutting the forest)

        3. *Biocentrism/Preservation*—don't use (e.g., No cutting the forest)

EARTH OVERVIEW

V. **Global coordinate system—network of east-west and north-south references used to record locations on the earth's surface**

    A. *Latitude*

        1. Measures north and south; northern and southern hemispheres

        2. Lines called parallels

        3. Equator = 0° latitude (no N/S)

        4. Northward to 90°N and southward to 90°S

        5. Low latitudes = 23.5°N–23.5°S

        6. Middle = 23.5°–55° N & S

        7. High latitudes = 55°–90° N & S

  B. *Longitude*

    1. Measures east and west; eastern and western hemispheres

    2. Lines called meridians

    3. Prime meridian = 0° longitude at Greenwich, England (no E/W)

    4. Eastward and westward to 180°—the International Date Line, in the mid-Pacific Ocean (no E/W)

VI. **Where people live on Earth**

  A. 90% of Earth's *7+ billion* people live:

    1. On 30% of the Earth's land surface

    2. In the mid-latitudes

    3. In the Northern Hemisphere

  B. Remaining 10% live in tropical forests, polar areas, deserts

  C. World's population increasing by ≥ 85 million per year; 2.5 people every second!

  D. What is Earth's **carrying capacity**—number of organisms that can live in a long term sustained balance with the environment at a reasonable quality of life

VII. **Humans on Earth**

  A. Have spread to all areas of earth

  B. Antarctica only continent w/o human activity and **land use**—any human activity that takes place on land

  C. Agriculture allowed increased **population density** (the number of people living in a geographical area divided by the total area of land) in some places

  D. Tend to live close to each other and on the best land

VIII. **Divisions of human uses of Earth**

  A. *Forests ~30% of total*

    1. Tropical—low latitudes

    2. Temperate—mid-latitudes

    3. Boreal—high latitudes

    4. 10% of total forest cover is gone!

    B.   *Grasslands* ~25%—most lost to agriculture

    C.   *Agriculture/Rural* ~11%–12%

        1.   ½ of world's farming population live on small, subsistence farms in **LDCs** (Less Developed Countries; 65% of population)—farming to support family with no surplus to sell or trade

        2.   **MDCs** (More Developed Countries; 35% of population)—large, commercial farms—most sold

    D.   *Cities/Urban*

        1.   70%–80% in MDCs live in cities

        2.   30%–40% (and growing) in LDCs

        3.   City sizes are increasing

        4.   People in MDC cities consume 2/3 of the Earth's natural resources.

    E.   Huge consumption of resources = massive amounts of waste

IX.   **Environmental spheres**

    A.   **Hydrosphere**—all the water in all forms (gas, liquid and solid [cryosphere])

    B.   **Lithosphere**—solid, inorganic (nonliving) materials

    C.   **Atmosphere**—the gaseous envelope that surrounds Earth

    D.   **Biosphere**—all living things (plants, animal, and microorganisms) on land and in the air and water

HISTORY OF AMERICAN CONSERVATION

X.   **American Indians—the continent's first conservationists**

    A.   Depended on nature for *everything*

    B.   Estimated 1–10 million in Americas

    C.   Specific culture determined by local environmental conditions

    D.   Small populations minimized environmental impacts

    E.   Are using their conservation techniques

XI.   **Early 1600s—Europeans settling East Coast**

    A.   Vast continent implied inexhaustible natural resources

    B.   Viewed continent as hostile wilderness

C. The colonists brought new ideas about land

1. Use of iron age technology

2. Humans could—and should—control their destiny

3. Concept of land ownership

D. New ideas resulted in new attitude and policies toward the land and its resources

E. At first, settlers had little negative impact except near settlements

F. 1700—all colonies had closed hunting season for deer; poorly enforced

G. Mid-1800s, some proposed government—owned wilderness be protected from exploitation

H. Late 1800s, overhunting and habitat destruction reduced many species' populations (video)

Chapter 2

The Hydrosphere

I. **Introduction to the Hydrosphere—water**

    A. Oceans cover 70% of Earth's surface

    B. Water changes phase—solid, liquid, and gaseous in nature

    C. *Energy absorbed* in the solid–liquid–gas change or *released* in the gas–liquid–solid changes

II. ***The Hydrologic Cycle***—continuous interchange of moisture from surface to atmosphere; no beginning or end

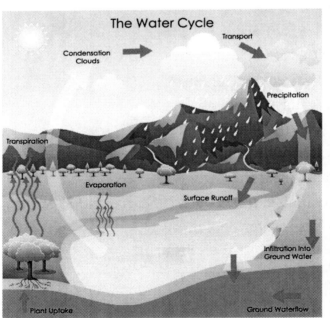

    A. Involves only 1% of total water

    B. *Components* of hydrologic cycle

        1. Evaporation *and* transpiration

        2. Condensation *and* cloud formation

3. Precipitation

4. Runoff *and* infiltration

III. **Natural reservoirs of water**

A. *Oceans—97.5%*, salty, cover 70% of Earth's surface

B. *Permanent glacial ice*

1. *2%* of total water, 75% of fresh water, covers 10% of Earth's surface

2. Types of glacial ice

a. Continental—ice sheet, ice caps

b. Oceanic—icebergs

c. Underground—**permafrost** - permanently frozen soil

C. *Underground*

1. 0.6% of total water

2. Hydrologic zones—surface downward

a. *soil water*

b. *Groundwater* (underground rivers)

c. *Aquifers* (underground lakes)

d. *Waterless zone*—too much pressure and rock density

D. *Surface*—0.01% of total

1. Wetlands (see Figures 2.8 [swamp] and 2.14 [marsh])

2. Rivers and streams (small river) (see Figure 2.6)

a. Drain water off land via a **watershed or drainage basin**-land area that contributes run-off to a river

3. **Lakes**—body of water surrounded by land—and ponds (small lake)

a. Reservoir—man-made lake

b. Most exist 100–1000 years

c. Conditions for a lake to form and persist:

(1) Natural basin or depression

(2) Restricted outflow of water

(3) Sufficient input of water

E.   *Atmospheric* water

1.   0.001% of total water on Earth

2.   Found naturally in all forms—solid (snow or hail), liquid (rain or fog), and gas (water vapor)

IV.   **Human activities affect waterways**

A.   Are withdrawing water at unsustainable rates

B.   Engineered freshwater waterways to meet human needs

1.   Water supplies (water wars in the future?)

a.   Unevenly distributed globally (see map Figure 2.20)

b.   Consumptive use: 70% to agriculture; 20% to industry

2.   Flood control

a.   Dikes and levees along river banks (see Figure 2.21)

b.   Dams originally built to prevent flooding

(1) Now also used for drinking water, irrigation, and electricity

(2) 45,000 Large ones ($\geq$ 49' high) across world in 140 nations

(3) Produce mix of costs and benefits (see Figures 2.22 and 2.23)

(4) 400 Obsolete dams in the United States removed and 500 more by 2020

V.   *Bottled water* **has ecological costs**

A.   2009—27 gals. of bottled water per American, costing $10.5 billion

B.   Energy costs of bottled water 2000× more than tap water

C.   ¾ Empty bottles (30–40 billion) into landfills

D.   Many tests show bottled water is no safer or healthier than most American's tap water.

E.   Read "Is it better in a bottle?" pp. 42–43

VI.   *Depletion* **of water reservoirs**

A.   Ground water depletion—*Ogallala aquifer*

1.   Under the Great Plains in the Midwest

2.   "Ancient water," from a wetter climate

3.   Rate water being pumped 20 times faster than natural recharge (see Figure 2.5)

    B. Surface water depletion—*Aral Sea* (see Figure 2.24)

        1. 1960–2009 area 90% less, volume 80% less

        2. Caused by irrigating 20 million acres of cotton from only two rivers feeding the Aral

        3. Too salty for fish to survive

        4. 40–50 million tons of salt onto croplands

        5. Colder winters and warmer summers

        6. Hope for the Aral

            a. In 2001, $85 million project began

            b. World Bank and Kazakh government reconstructed 60 miles of canals/waterworks along the Syr Darya, doubling its flow rate into the North (Small) Aral Sea.

            c. Centerpiece of project—Kok-Aral Dam, 8 miles long, keeping the water from South Aral (Large) Sea.

            d. Since completion in August 2005, North Aral (Small) Sea increased in area by 18%, covering >300 mi^2 of parched seabed, and depth by 2 miles by 2007, which allowed fishing again.

VII. ***Water Quality* in the United States**

    A. Major water quality acts

        1. 1899—Refuse Act—first federal actions to control water pollution

        2. 1924—Oil Pollution Control Act (revised 1990)

        3. 1948—Water Pollution Control Act

        4. 1972—*Clean Water Act*

            a. Set federal water quality standards

            b. Federal grants to states to build or improve waste water treatment plants

    B. Environmental Protection Agency (EPA)—enforcer of acts

VIII. ***Sources* of water pollution—addition of harmful or objectionable materials that causes a reduction in water quality**

    A. *General types* of sources (see Figure 2.25)

        1. *Point source*—defined origin(factory)

        2. *Nonpoint source*—unclear origin(farmland)

B. *Specific* sources—*where* they come from

1. *Agriculture* (nonpoint)

    a. Produces sediment, manure, fertilizer, pesticides

    b. World's largest nonpoint source

2. *Urban environments*

    a. Wastewater sewage treatment (point)

    b. Creates sludge—toxic materials removed from wastewater

3. *Storm water runoff* (nonpoint)

    a. Carries debris

    b. Amount increases with **land use density**—percent of land covered by impervious surface

4. *Soil erosion* (nonpoint)

    a. Land clearing for development and agriculture

5. *Atmosphere*—air pollution

    a. Power plants (point)

    b. Vehicles (nonpoint)

IX. **Specific forms—*what* they are**

1. *Toxic chemicals*

    a. Arsenic, lead, mercury, acids

    b. Pesticides, petroleum products, and so on

2. *Disease-causing organisms*

    a. Mostly from inadequately treated human or animal waste water

    b. 1 billion people do not have safe drinking water (see Figure 2.20)

    c. Kills 5 million per year worldwide

3. *Oil pollution* (see Figure 2.26)

    a. 2010 BP Deep Horizon explosion; 1800 gpm

    b. Natural seepage from ocean floor (47%)

    c. Widely-spread small nonpoint sources (38%)

    d. Spills during transport (12%)

    e. Leakage from offshore oil extraction (3%)

    f.   US Oil Pollution Act of 1990 created $1 billion prevention and clean-up fund

    g.   By 2015, all oil tankers required to have double hulls

4.   *Nutrients and* **eutrophication**-excess nutrient enrichment in water bodies

    a.   Addition of nutrients (esp. phosphorus and nitrogen from waste) promotes excess *algae* growth on surface of water bodies

    b.   Algae blocks sunlight to plants on bottom, and *consumes oxygen in the water* when dies/decomposes (see Figure 2.27)

5.   *Sediment*

    a.   Rock and mineral fragments of various sizes

    b.   By *volume and mass*, is *greatest* water pollutant

    c.   Impacts from soil erosion

        (1)   Depletes soil at site of origin

        (2)   Reduces quality of water resource it enters

            (a)   Impairs the respiration of fish and invertebrates, and smothers bottom dwelling organisms

            (b)   Clouds water that blocks sunlight to bottom plants

6.   *Thermal pollution*

    a.   Warm water holds less dissolved oxygen than cool water

    b.   Use water to cool facility; heated water returned to stream

    c.   Removing stream bank vegetation increases water temperature

7.   *Nets and plastic debris* (READ p. 45, *Envision-it*)

    a.   Fishing nets, plastic bottles and bags, fishing line, buckets, floats, and so on

    b.   Aquatic mammals, seabirds, fish and sea turtles mistake floating plastic debris for food; kills them

    c.   Collects where ocean currents converge

        (1)   "The Great Pacific Garbage Patch"

        (2)   Twice size of TX

    d.   3.5 plastic bits/m^2 in water

8. **Acid mine drainage**—acidic water that flows from underground mines

   a. Coal associated with iron and metal sulfides

   b. When comes in contact with oxygen and water, breaks down into smaller pieces (weathers)

   c. Produces sulfuric acid when water runs through or over the sulfides

   d. Acidic water toxic to aquatic plant and animals, and is water pollutant

   e. Significant problem in several states, esp. WV, MD, PA, OH, and CO

X.   *How much available, fresh water?! Scarce and Precious!*

Double Brain/Shutterstock.com

XI.   *Solutions* to fresh water depletion

A. Build more dams—already built at suitable sites

B. Desalinization—requires large amounts of fossil fuels

C. *DECREASE OUR DEMAND FOR WATER*

   1. Agriculture—Improve irrigation methods—drip or down-facing sprinklers

   2. Households—Install low-flow, high-efficiency toilets, showerheads, washing machines, and dishwashers or Plant native vegetation

   3. Industry and municipalities – recycle water; fix leaks

XII.   **Water conservation**

    A.   *75%* of household water is used to flush toilets and bathe

        1.   Ten minute shower = 50 gallons

        2.   Each toilet flush = 5 gallons

    B.   Use of conserving fixtures saves a lot

        1.   Water-conserving shower head = 25 gallons

        2.   low-flow toilets = 1.5 gallons

Chapter 3

Geology, Minerals, and Mining

I. **Vertical Structure of Earth's Interior** (see Figure 3.1)

    A. *Crust*

        1. Thinnest layer

        2. Oceanic—5–6 miles deep; continental—15–20 miles deep

        3. MoHo—transition zone between the crust and the mantle

    B. *Mantle*

        1. Thickest layer; ¾ total volume

        2. *3 distinct zones* within mantle

            a. *Lithosphere*—includes crust and top of mantle; solid rock

            b. *Asthenosphere*—upper mantle

                (1) Extends 200 miles down

                (2) Made of "liquid" rock

                (3) **Lithospheric plates** reside there—huge sheets of the lithosphere that are imbedded in the nonrigid asthenosphere that shift extremely slowly across Earth's surface over millions of years

                (4) Movement of the plates is due to **plate tectonics**—geological theory that describes the process that Earth's crust shifts slowly, moving continents and reshaping ocean basins (see Figure 3.2)

                (5) Results of plate movement is **continental drift**—theory of the movement of land masses and oceans over the past 225 million years (my) to their current location via plate tectonics (see Figure 3.2)

                (6) Plate tectonics and continental drift helped explain the location of some geologic features

                    (a) Mountains and volcanoes—plates moving together; for example: Mt St. Helens (see Figure 3.5)

                    (b) Earthquakes—plates moving apart or sliding past each other; for example: San Andreas Fault

                    (c) Oceanic islands—"hot spots"; for example: Hawaiian Islands (see Figure 5.7)

            c. *Mesosphere*—lower mantle; 1800 miles down; solid rock

    C. *Outer core*—liquid rock, 3200 miles deep

  D. *Inner core*—solid and dense

   1. Radius of 700–750 miles

   2. Made of iron, silicate and nickel

  E. *Radius* of Earth ~4000 miles

II. Crustal composition

Relative percentages by weight of chemical elements in the Earth's lithosphere	
Element	**% Mass**
gas Oxygen (O)	45.2
Silicon (Si)	27.2
Aluminum (Al)	8.0
Iron (Fe)	5.8
Calcium (Ca)	5.1
Magnesium (Mg)	2.8
Sodium (Na)	2.3
Potassium (K)	1.7
Titanium (Ti)	0.8
gas Hydrogen (H)	0.8
Manganese (Mn)	0.1
All others	0.1

99.9%

  A. 90 chemical elements bond to form compounds

  B. Bonded chemicals = *minerals*

   1. Minerals must:

    a. Be found in nature

    b. Be composed entirely of inorganic (nonliving) material

    c. Have the same chemical composition everywhere

    d. Have atoms arranged to form solid units called crystals

   2. 4000 know minerals, some extraterrestrial

   3. *Only 20* of 4000 minerals make up 95% of all **rocks**—a collection of minerals

      4.   **Gem**—a rock of a single mineral formed very deep, under very high pressure and where no impurities are present (e.g., diamonds are pure carbon)

C. Rocks

    1.   **Bedrock**—residual rock that has not been exposed to erosion

    2.   **Regolith**—a layer of broken rock particles that covers bedrock

    3.   Rock type #1 = *Igneous rock*

        a.   Formed by the cooling and solidification of molten rock

        b.   **Magma**—molten rock *under* the earth' surface

        c.   **Lava**—molten rock once it *reaches the surface* where it cools and solidifies

        d.   **Pyroclastics**—solid rock fragments thrown into air by volcanic explosions; can fuse together to *form igneous rock*

    4.   Rock type #2 = *Sedimentary rock*

        a.   Formed of sediment consolidated by the combination of pressure and cementation

        b.   Sediments transported by moving water are eventually deposited in a quiet body of water (e.g., ocean floor)

        c.   Weight of overburden creates enormous pressure which causes *compaction* of individual particles

        d.   Cementing agents from water go into pore space causing *cementation* of the particles

    5.   Rock type #3 = *Metamorphic rock*

        a.   Was originally igneous or sedimentary rock but has been drastically changed by massive forces of heat and/or pressure from within the earth.

        b.   Geologic forces may bend, uplift, compress, or stretch rock

        c.   Some rock types always become another rock type when metamorphosed

            (1) Limestone becomes marble

            (2) Sandstone becomes quartzite

            (3) Shale becomes slate

        d. When metamorphosed, many other rocks become either *gneiss or schist;* too deformed to identify what the original rock was

    6. **Rock cycle**—the extremely slow process in which rocks and minerals are heated, melted, cooled, broken, and reassembled forming one of the three rock types (see Figure 3.4)

  D. Mineral location and classification (see Figure 3.13)

    1. **Ore deposits**—where minerals are concentrated by geologic processes

        a. From which individual minerals are extracted

        b. Minerals in crust are not evenly distributed because geologic and biologic processes selectively dissolve, transport, and deposit them

            (1) Metallic ores deposited in the crust where lithospheric plates move

            (2) Ore deposits form when molten rock cools

            (3) Processes of transporting sediments by wind, water, and glaciers often concentrate minerals in certain deposits

            (4) Minerals are concentrated in soil by **weathering**—the chemical and mechanical decomposition (break down) of rock

    2. *Mineral resources vs mineral reserves*

        a. **Mineral resources**—minerals and rocks concentrated in a form that *could be* extracted to obtain a useable commodity

        b. **Mineral reserves**—known and identified deposits of the resource that *can be* extracted profitably, with existing technology and under current economic and legal conditions

        c. Whether a mineral deposit is classified as part of the *resource base* or as a *reserve* is often determined by economics

            (1) If a particular mineral became scarce, the price would increase and encourage exploration and mining

            (2) Resulting from the price increase, previously uneconomic deposit (part of the *resource base*) may become profitable and would be re-classified as mineral *reserve*

3. Two *types* of minerals

    a. *Metallic minerals* and subtypes

        (1) *Abundant*—common and easily accessible to humans; for example: iron, aluminum, magnesium, titanium, manganese

        (2) *Scarce*—are NOT common and NOT easily accessible to humans; for example: copper, lead, zinc, nickel, gold, silver

assistant/Shutterstock.com

    b. *Nonmetallic minerals* and subtypes

        (1) Represent the natural resource used in the *greatest volume*

        (2) Building materials; for example: stone, sand, gravel, cement, plaster

Dmitri Melnik/Shutterstock.com

        (3) In fertilizer and chemical products; for example: nitrogen, phosphorus, potassium, halite (salt), sulfur

III.   Mining techniques

    A.   Placer mining (see Figure 3.17)

        1.   Technique involving sifting through materials in riverbed deposits, using running water to separate lighter mud and gravel from heavier minerals of value

        2.   Environmentally destructive—large amounts of sediment and debris washed into streams which makes them uninhabitable for aquatic organisms for many miles downstream

    B.   Solution mining

        1.   Technique involving bore holes drilled deep into the ground to the deposit and water, acid or other liquid injected into the bore holes to dissolve into the liquid

        2.   Less negative environmental impacts than most mining techniques, but accidental leakage of the liquids into groundwater can cause contamination

    C.   Strip mining (see Figure 3.14a)

        1.   Layers of surface soil and rock (overburden) are removed from large areas to expose minerals by giant machinery and placed in *spoil banks*

        2.   Mineral is removed from that strip of land and the overburden is put back into the space before moving on to another strip of land

        3.   Technique used mainly for coal, but sometimes sand and gravel

    D.   Open pit mining (see Figure 3.16)

        1.   Removal of materials by digging them out of the earth and leaving a large pit

        2.   Most suited for areas where the mineral is spread widely and evenly through the rock formation

        3.   Technique used for mining copper, iron, gold, diamonds, and coal

        4.   Clay, gravel, sand and stones (e.g., limestone, granite, marble, slate) mined by same technique but called quarries

    E.   *Negative impacts* of strip and open pit mining

        1.   Overburden piled in spoil banks highly erodible and eroded chemicals and sediments enter our water systems

        2.   Pollutes the air with dust and fumes

        3.   Animals and plants displaced or destroyed

        4.   Makes the landscape ugly and unusable!

F. **Subsurface mining** (see Figure 3.14b)

1. Shafts dug deeply into the ground and tunnels dug and blasted to get to the coal

2. *Most dangerous* technique for miners

    a. Cause collapse (Centralia, PA)

    b. Natural gas explosions

    c. Inhale toxic fumes and coal dust (Black Lung disease)

3. *Environmental hazards* include groundwater contamination due to depth of shaft, and surface water contamination due to run-off flowing through mines and creating acid mine drainage

G. *Surface Mining Control and Reclamation Act, 1977*

1. Act requires mining companies to restore a mining site close to its premining condition (see Figure 3.19)

2. Restoration involves:

    a. Remove buildings and other structures used for mining

    b. Fill in area and grade to original contour of land

    c. Fill in shafts (underground)

    d. Replace topsoil

    e. Plant fast-growing vegetation to keep soil in place

3. Costs $1000–$5000 per acre

4. Success if variable, especially in arid and semiarid regions

IV. **How to reduce raw mineral extraction and its negative impacts**

A. *Conserve*—use what we have longer (see Figure 3.20)

B. *Recycle*—advantages

1. Extends supply of nonrenewable minerals (see Figure 3.21)

2. Uses less energy and is less expensive than mining raw minerals

3. Causes less air and water pollution and less land disruption and degradation

4. Prolongs the lifetime of landfills

   C. *Substitute* new materials for the minerals

      1. Copper, lead, and steel pipes to PVC

      2. Copper telephone wires to fiber optics

      3. Steel in automobiles changed to polymers, aluminum, and ceramics, making them lighter and therefore more fuel efficient

   D. *Issues* related to substitution

      1. Finding and developing a substitute that is as good as or better than the original mineral is expensive and time-consuming

      2. Phasing the substitute into the manufacturing process will take time and cost money

      3. Will lead to an increase in production costs = higher cost for consumer

      4. Even with issues, still worth it in the long run!

Chapter 4

Soil, Agriculture, and Food

I.   Can we feed the world?

    A. Development of agriculture enabled human population to grow rapidly

    B. As we improved agricultural conditions, there has been a constant effort to overcome environmental limitations and problems.

        1. Each solution to these problems has brought their own set of environmental problems and *they* require new solutions

        2. The cycle is unavoidable and must be acknowledged and addressed continually

    C. As population increases, so does the need to grow more and more food

        1. Future food production will need to be increased per unit area (more food in same or less space)

        2. "Best" land already being farmed, but is under constant pressure to be converted to other uses (farmland preservation)

        3. Land conversions result in having to farm increasingly marginal land, which usually means an increased need for more water and fertilizer

II.   What we grow and eat

    A. Approximately 3000 crop species grown worldwide; ~200 in United States

    B. 150 crops worldwide grown on *large scale*

    C. *Just 14 supply most of the world's food*—wheat, rice, maize (corn), potatoes, sweet potatoes, manioc, sugarcane, sugar beets, common bean, soybeans, barley, sorghum, coconuts, and bananas

    D. Also grow forage for animals to eat, especially alfalfa, sorghum and hay

MaraZe/Shutterstock.com

E. **Aquaculture**—production of food from aquatic habitats (see Figure 4.25)

　　1. Growing worldwide

　　2. Example: carp, tilapia, oysters, shrimp (crayfish, catfish, and trout also in United States)

F. **Mariculture**—production of food from ocean habitats

　　1. Growing worldwide; very productive

　　2. Grown on rafts in ocean or on artificial pilings in the tidal zone

　　3. Oysters and mussels obtain food from water moving past with the tides

G. *Historical stages* of agriculture

　　1. *Traditional, resources-based* (based on conservation of land, water, and energy) and **organic farming** (agriculture that uses no chemical fertilizers or pesticides, relying on biological approaches) was introduced 10,000 years ago and was mostly **polyculture**—planting multiple crops in a mixed arrangement or in close proximity

　　2. *Industrial, demand-based* (based on highly mechanized technology requiring land, water, and fuel resources) arose during the Industrial Revolution, 18th and 19th centuries; typically **monoculture**—the planting of a single crop over large areas

　　3. Return to *resource-based* in 20th century, using new technologies

　　4. Currently, increasing interest in organic farming technologies (see Figure 20.26) as well as the opposite, **genetic engineering**—any process scientists use to manipulate an organism's genetic material in the lab by adding, deleting, or changing segments of its DNA

H. **Genetically-modified crops (GMCs)**—crop species modified by genetic engineering to produce desired characteristics or eliminate undesirable characteristics in organisms (see Figures 4.18 and 4.19)

　　1. Have introduced new environmental debates, but have the potential for increasing food production on the same or less space

　　2. Three practices use GMCs

　　　　a. Faster, more efficient ways to develop **hybrids**—offspring of genetically dissimilar parents, especially offspring produced by breeding plants or animals of different species or varieties

　　　　　　(1) Hybridization standard methodology in agriculture (mule)

　　　　　　(2) New ways to create hybrids by **biotechnology**—application of the use of genetic engineering techniques

    b. Introduction of the "terminator gene"

      (1) Makes seeds from crop sterile

      (2) Meant to keep GMCs from spreading into environment

      (3) Small-scale farmers usually use this year's seeds for next year's crops; can't do that

    c. Transfer of genetic properties

      (1) Allow crops to "fix" nitrogen (make available to crop)

      (2) To resist herbicides so farmers can apply them to kill weeds without killing the crop

      (3) To resist insects, drought, cold, heat, toxins, disease

3. In three decades GMCs have gone from science fiction to mainstream agriculture

4. Allows large-scale commercial farmers to grow crops more efficiently

5. In most of the world, soybeans are the GM crop grown the most (93%); corn is second (86%) (see Figure 4.20a)

6. The United States grows 45% of all GM crops (see Figure 4.20b)

III. Soil—the lithosphere's second natural resource

A. A relatively thin, natural part of Earth's surface that is a porous layer of mineral and organic matter, and which is formed as the result of chemical and biological processes on rocks over time; produces and stores plant nutrients

Zigzag Mountain Art/Shutterstock.com

    B. How much of the world's land can grow food? (demonstration)

    C. Soil components

        1. *Inorganic* material—45%; ½ small pieces of weathered rock fragments and ½ small particles of minerals

        2. *Organic* material—5%

            a. Living and dead plants and animals

            b. Makes soil be more than dirt by giving it life (see Figure 4.5).

            c. **Humus**—dark soil matter composed of decayed organisms

                (1) The end result of plant decomposition

                (2) Improves soil fertility

        3. *Soil air*—found within the *pore space* between soil particles

            a. Always saturated because surrounded by water

            b. Rich in carbon dioxide and poor in oxygen

        4. Soil water

            a. Most gets into soil by infiltration of precipitation

            b. Quantity variable and depends on:

                (1) Temperature and precipitation

                (2) Amount of vegetation

                (3) Characteristics of the soil such as *texture* (particle size)

                    (a) *Sand*—large grains; does not stick together well; water passes through easily

                    (b) *Silt*—medium-sized grains; best for plant growth

                    (c) *Clay*—very small grains; sticks together easily; does not allow water to pass through easily (absorbs it)

            c. *Performs* important functions

                (1) Dissolves plant nutrients, making them available to plant roots

                (2) Enables chemical reactions, for example: production of humus

                (3) **Leaching**—plant nutrients carried below level of plant roots due to too much precipitation or irrigation; bad for plants

IV.   Soil management and conservation

A. **Soil degradation**—a deterioration of soil quality and decline in soil productivity

   1.   Yearly, gain > 80 million people globally, but lose 12–17 million acres (WV-sized) of productive cropland to degradation

   2.   Most common causes

        a.   Soil erosion, chemical pollution, nutrient depletion, water scarcity

        b.   Salinization, waterlogging, loss of organic matter, changes in soil structure or soil chemistry

B. **TOPIC 1—Soil erosion**—detachment and removal of soil by water, wind and/or gravity

   1.   A natural process from movement of wind and water

   2.   Vegetation holds soil in place and reduces erosion

   3.   Land more vulnerable to erosion because of:

        a.   Over-cultivating fields through poor planning or excessive tilling

        b.   Overgrazing of rangeland

        c.   Clearing of forests on steep slopes or clear cuts

   4.   Water erosion occurs in a *sequence from* very small areas to very large areas

        a.   **Rainsplash (drop) erosion**—the direct collision of a raindrop with the ground which blasts fine particles upward and outward, shifting them a fraction of an inch

        b.   **Sheet erosion**—the transportation of material loosened by rainsplash, and is accomplished by water flowing across the surface as a thin sheet; sometimes called *overland flow* (see Figure 4.8)

        c.   **Rill erosion**—a more concentrated flow of water than sheet erosion which loosens additional material and causes parallel, shallow flows on a slope

        d.   **Gully erosion**—overland flow that erodes conspicuous channels in the soil; destroys the land

   5.   Wind erosion occurs the most in dry climates on flat land (see Figure 4.9)

        a.   Amount of soil lost by wind erosion may not be realized because the soil particles disperse over large areas where they are not visible

b.   Soil eroded by wind can travel extremely long distances

c.   The impact of bouncing soil particles dislodges finer particles and put them in suspension; coarser particles move by surface creep

C.   Common methods to *minimize soil erosion*

1.   **Crop rotation**—a series of different crops planted in the same field in different years; helps replace depleted soil nutrients (see Figure 4.10a)

2.   Reducing water erosion on *slopes*

a.   **Strip-cropping**—planting different species of crops in alternating strips along land contours, and **contour plowing**—plowing across a slope rather than up and down it—if used together are very efficient (see Figure 4.10b)

b.   **Terracing**—shaping the land to create level shelves of earth to hold water and soil; requires extensive hand labor or expensive machinery, but allows farming on very steep slopes (see Figure 4.10c)

3.   Use of *ground cover* to minimize *water* erosion

a.   **Intercropping**—the planting of different types of crops in alternating bands or other spatially mixed arrangements; slows erosion by covering more ground than a single crop (see Figure 4.10d)

b.   Crop residue left in field (e.g., corn stalks from previous harvest)

c.   Planting cover crops after harvest holds soil in place in winter and can be plowed under in spring to add organic matter to soil

d.   Mulch reduces erosion and saves water, adds nutrient as it decomposes, retards weed growth, wide variety of possible materials to use

4.   Use of *ground cover* to minimize *wind* erosion by planting **windbreaks or shelterbelts**—rows of trees and shrubs arranged at right angles to the prevalent wind direction to diminish erosion on crop land (see Figure 4.10e)

5.   *Reduced tillage* methods

a.   **Minimum till**—use of a chisel plow that slices a trench where seeds are planted rather than a traditional plow

b.   **Conservation till**—use of a sharp disk that slices the soil just wide enough to insert the seed

c.   **No-till**—a plow is not used, but rather a special planter that punches a hole in the soil to put the seeds (see Figures 4.10f and 4.11)

D. **TOPIC 2—Irrigation**—the artificial provision of water to support agriculture

   1. *Problems* with irrigation

      a. *Poor efficiency* due to evaporation and seepage

      b. *Overwatering* wastes water and can cause **waterlogging**—saturation of plant root zone with water; no air; drowns plants

      c. **Salinization**—after-effect of irrigating land with poor drainage in which evaporation of salty water leads to salt accumulation at the surface

      d. *Increased erosion* causing runoff of chemicals and sediment

      e. *Depletes aquifers* causing land subsidence

   2. Irrigation *techniques*

      a. *Flood the field*—most common but least efficient due to much water lost to evaporation and seepage

      b. *Sprinklers*—some are ineffective due to water being lost to evaporation (see Figure 4.13a); central pivot sprinklers best

      c. *Drip or trickle*—water applied to the soil from tubes that drip water slowly onto each plant; best method (see Figure 4.13b)

         (1) Water delivered directly to plant root (liquid fertilizer, too)

         (2) Minimal loss to evaporation

         (3) Overwatering and runoff prevented

         (4) Expensive initial investment but worth it in long run

E. TOPIC 3 Soil Fertility

   1. To obtain good soil fertility and therefore good plant growth, the soil needs

      a. First good organic content

      b. Second proper pH level—pH 6 is 10 times more acidic than pH 7

         (1) Most crops prefer a neutral soil pH

         (2) Lime is used to increase pH (decrease acidity)

      c. Third sufficient nutrients, especially nitrogen, phosphorus, potassium

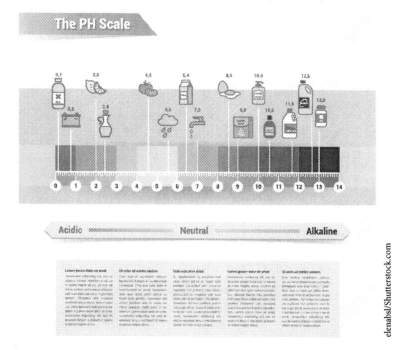

elenabsl/Shutterstock.com

2. Inorganic fertilizer—synthetically manufactured mineral supplements

    a. Most used since 1900s

    b. Boosted food production worldwide

    c. Overapplication causes air and water pollution problems

3. Organic fertilizer—made of natural materials

F. TOPIC 4 Pesticides—a general term

    1. **Pest**—any organism that reduces the availability, quality or value of a resource useful to humans; vary from place to place

    2. **Pesticide**—any chemical that kills, controls, drives away or modifies the behavior of a pest

    3. Specific pesticide identified by its target organism (e.g., insecticide)

    4. Are absorbed by and bound to soil particles; not passive or temporary

    5. *Benefits* of chemical pesticides

        a. Increased food production by 50%

        b. Decreased disease carried by insects, especially mosquitos

    6. *Problems* with pesticides

        a. Pests often develop a resistance to pesticides over time (see Figure 4.15)

    b. Runoff carrying pesticides pollute water supplies

    c. Poisonous to humans

7. *Most used* pesticides

    a. *Herbicides*—weed killers; 55% of all pesticides; <10% reach target weeds

    b. *Insecticides*—bug killers; 25% of all pesticides; *more toxic* to humans than herbicides

    c. *Fungicides*—fungus and mildew killers; 15% of all pesticides

        (1) Protects crops from fungal disease

        (2) Major use—to prevent spoilage of fresh fruits and vegetables

        (3) *Most toxic* to humans; WASH THEM!

chinahbzyg/Shutterstock.com

8. Techniques in *biological insect control*

    a. Use of "natural" insecticides found in some plants (marigolds, mums)

    b. Use of a harmless insect to control a pest insect

    c. Insect traps—Bag-A-Bug

    d. Interference with reproduction in extreme situations

Chapter 5

Atmospheric Science and Air Pollution

OUTLINE

    I. Introduction to the Atmosphere

   II. Outdoor air pollution (OAP)

  III. Acid rain and deposition

  IV. Indoor air pollution (IAP)

   V. Stratospheric ozone depletion

I.   **Introduction to the Atmosphere**

   A.   Life on Earth because of the abundance of water, just right amount of solar radiation and our atmosphere

   B.   Air supplies oxygen to animals, carbon dioxide to plants, recycles our water, reduces surface temperature extremes, shields us from UV rays

   C.   *Components* of air

   1.   *Gases*—two types (see Figure 5.1)

   a.   **Nonvariable**—volume in atmosphere *does not* change over time or space

   (1)   *Nitrogen (78%)*

   (2)   *Oxygen (21%)*

   (3)   *Argon (0.09%)*

   b.   **Variable**—volume in atmosphere *does* change over time and/or space

   (1)   *Water vapor*—gaseous form of water; varies 0%–4% in volume

   (2)   *Ozone*—in stratosphere; absorbs UV rays; decreasing

   (3)   *Carbon dioxide* (CO_2)—increasing

   (a)   Used to think it was nonvariable

   (b)   Increasing due to fossil fuel burning and deforestation

   2.   **Particulates**—microscopic solid particles that are suspended in the air; major ones = dust, smoke, salt, pollen

   D.   Atmosphere's *vertical structure*

   1.   **Atmospheric pressure**—the force applied to the surface by the weight of overlying air

   a.   Highest at the surface

   b.   Decreases upward at a *decreasing* rate

   E.   *Temperature* (see Figure 5.2)

   1.   *Troposphere*

   a.   Begins at surface

   b.   *Decrease* of temperature with height

   c.   Clouds and weather there

   d.   **Tropopause**—transition zone with *no change* of temperature with height for a few miles

2. Stratosphere

   a. **Temperature inversion**—an *increase* of temperature with height in the atmosphere

   b. Inversion exists because of the "good" ozone there absorbs heat energy

   c. Stratopause

3. *Mesosphere*

   a. *Decrease* of temperature with height

   b. Mesopause

4. *Thermosphere*

   a. Temperature inversion because of very thin air there

   b. No thermopause; gradually becomes outer space

II. Outdoor air pollution (OAP)

A. Introduction

   1. Mixing and reacting chemicals in the troposphere create potential **air pollutants**—airborne particles and gases that *occur in harmful concentrations* that endanger health and the well-being of organisms, or disrupt the orderly functioning of the environment

   2. 100s of air pollutants, but considered a MAJOR pollutant if:

      a. It is known to have a significant negative impact on human health and/or the environment, AND

      b. It is produced by human activities, in large quantities, over much of the earth

B. *Classes* of air pollutants

   1. **Primary**—a chemical that has been *added directly* to the air by natural events or human activities and *occurs in harmful concentrations*

   2. **Secondary**—a harmful chemical *formed in the atmosphere* by reacting with normal air components (like oxygen) or a primary air pollutant

C. *Types* of outdoor primary air pollutants (see Figure 5.8)

   1. *Carbon monoxide*—CO; from incomplete fuel combustion

   2. *Nitrogen oxides*—NO_x; from reaction of N_2 and O_2 due to high temperature engine

3. *Volatile organic compounds*—VOCs; from industrial solvents and vehicles

4. *Sulfur dioxide*—SO$_2$; from electricity and industry

5. *Particulate matter*—PM$_{2.5}$ and PM$_{10}$; from dust and combustion processes

D. *Sources* of air pollution

   1. *General categories* of air pollution sources

      a. Natural (see Figure 5.6)

      b. Stationary

      c. Mobile (see Figure 5.11)

   2. Major *specific sources* of air pollution

      a. Transportation *(mobile)*

      b. Fuel combustion from *stationary* sources

      c. Industrial processes *(stationary)*

      d. Solid waste disposal *(stationary)*

      e. Miscellaneous, including *natural* (e.g., fire)

E. Smog: our most common air quality problem

   1. Originally meant *smoke and fog*

trekandshoot/Shutterstock.com

2. Now two *types* from different sources:

    a. **Industrial smog**—gray-air smog caused by the incomplete combustion of coal and oil when burned (see Figure 5.12a)

    b. **Photochemical smog**—brown-air smog caused by light-driven reactions of primary pollutants with normal atmospheric compounds that produce a mix of chemicals, especially "bad" ozone (see Figures 5.13 and 5.14)

3. *Steps in the formation* of photochemical smog

    a. Step 1 Morning rush hour traffic emit *nitric oxide* (NO) from their tailpipes

    b. Step 2 *Nitric oxide* reacts with *oxygen*, forming *nitrogen dioxide* by mid-morning

    c. Step 3 *Ultraviolet radiation* causes various chemical reactions with the *nitrogen dioxide*

    d. Over time, these chemical reactions create components of *photochemical smog* (e.g., "bad" ozone)

4. *Influences* on the severity and frequency of smog

    a. Climate

    b. Population and industry density

    c. Major fuels burned (coal vs nuclear)

    d. Vertical temperature structure of the troposphere

    e. Topography (valleys hold it)

F. Outdoor air pollution history

1. *World's first* air pollution disaster

    a. Meuse Valley, Belgium, December, 1930

    b. 63 deaths; 6000 ill

2. *First United States* air pollution disaster

    a. Donora, PA, October 1948 (see Figure 5.12)

        (1) Steel mill, zinc smelter, and sulfuric acid plant in city

        (2) 21 deaths; ½ town (6000) hospitalized

    b. *World's worst* air pollution disaster

        (1) London, England, December, 1952

        (2) 4000 died that month; 8000 more in January and February 1953

G. *Controlling* outdoor air pollution in the United States

1. *Clean Air Act,* 1970 and 1977—gave the federal government authority to control air pollution; amended 1990 (see Figure 5.9a)

2. *Environmental Protection Agency* (EPA) established:

   a. *Standards* that specify the maximum allowable levels of each air pollutant (see Figure 5.9)

   b. A *"policy of prevention"*—can't move polluting factories to clean air areas to escape clean air laws (still today)

   c. *Enforcement* of the standards

3. Each vehicle pollutes much less than in the past, but we have increased the number of miles driven and the number of vehicles on the roads, basically negating the advances we made by improving each vehicle's emissions (see Figure 5.9b)

III.   Acid rain and deposition

A. *Formation*

1. As water evaporates, dissolved and suspended substances (e.g., salt) are left behind at the surface or carried away by wind

2. Falling solid and liquid precipitation collects the human-produced and natural air pollutants, *changing the precipitation's chemistry*

3. *Natural* precipitation dissolves atmospheric *carbon dioxide*

    a. Produces a weak *carbonic acid solution*

    b. Results in *average* pH of 5–5.6

    c. pH ≤ 5.0 = harmful effects to environment (see Figure 5.19)

4. Acid rain caused by air pollution (see Figure 5.18)

    a. Electrical and industrial power plants that burn coal and oil, emit large amounts of sulfur dioxide (SO_2) and nitrogen oxides (NO_x) from smokestack into air (NO_x from transportation, too)

    b. Chemical reactions in the air cause SO_2 *emissions to become sulfuric acid* and NO_x *emissions to become nitric acid*, the two major components of acid rain

    c. The longer the emissions remain in the air, the more likely necessary reactions will occur and the acids will be formed

5. Chemicals reach ground two ways:

    a. Liquid acid precipitation as rain, snow, fog, dew, clouds

    b. Dry solid acid deposition fall as dry particles or the gases are absorbed by particulates

        (1) Dry gases become acidic as soon as interact with any moisture

        (2) Secondary "dry" acid has negative effects equivalent to primary "wet" acid

B. *Location* of acid problems

    1. Polluted emissions often carried long distance from sources by global wind

        a. All emitters are within the *westerly wind system*

        b. Therefore pollutants carried E/NE from their sources (e.g., Midwest) before falling as acid precipitation (e.g., New England and S. Canada) (see Figure 5.19)

    2. This long-range transport of resultant acid rain transcends political boundaries between the producing and receiving locations

C. Role of *geology*'s impact on acid rain

    1. Areas with *limestone or chalk* bedrock and soils can *"buffer"* (neutralize) additional acid input

    2. Areas with *granite or quartzite* bedrock and soil *cannot buffer* addition acid input, causing the water and soil to become acidic

D. *Effects* of acid rain on *aquatic environments*

1. Age and species composition of fish populations change

2. Toxic metals leached from soil and rock enter water systems

3. Food for higher lifeforms (e.g., insects) negatively affected by acid and toxins

4. "Spring shock" can overwhelm buffers

E. *Effects* of acid rain on *terrestrial environments*

1. Toxic metals leached out of soil and rock damage plant roots

2. Trees more susceptible to stress (e.g., drought, insects, disease)

3. Kills trees by (see Table 5.1)

a. Carrying nutrients away from roots

b. Inhibit photosynthesis

c. Physical damage to leaves or needles

4. Stunts the growth of many food crops

F. *Effects* of acid rain on *human environments*

1. Many human-built structures made of *limestone* which is susceptible to acid rain attack

2. Damage is expensive to repair and may be dangerous

3. World's culture at stake; have withstood war and severe weather but not acid rain

G. *Effects* of acid rain on *human health*

1. A *direct* effect—elevated atmospheric acidity major cause of respiratory problems in eastern North America due to sulfuric acid being inhaled

2. An *indirect* effect—leached toxic metals from soil and rock, as well as from pipes and tanks, can enter our water supplies or can be within the plants and animals humans eat

H. Possible *solutions to* the acid rain problem

1. *Initial step*—reduce SO_2 and NO_x emissions!

2. To *treat symptoms* of acid rain in lakes, lime can be applied

a. Sweden limes its 1000 lakes since 1970s, repeating every 3–5 years; temporary fix

b. Liming fields with lime successful in reducing acidity in agricultural fields

3. Best to tackle the problem at the *source* rather than treat the symptoms

4. *Three general approaches* to reducing industrial SO_2 output to air

   a. BEFORE the combustion of fuel

      (1) *Fuel switching*—use other fossil fuel or better quality of coal

      (2) *Fuel desulphurization*—coal washing or gasification

   b. DURING the combustion of fuel

      (1) Burning coal *with* lime

      (2) Only ones to reduce NO_x as well as SO_2

      (3) *Lime injected multistage burning technology (LIMB)*

      (4) *Fluidized bed combustion (FBC)*

   c. AFTER the combustion of fuel

      (1) *Flue gas desulphurization (FGD)* (see Figure 5.10)

      (2) Uses scrubbers to neutralize gases from smokestacks

         (a) Dry—filters smoke through lime or charcoal

         (b) Wet—gases passed through an alkaline agent, neutralizing the gases

         (c) Can be added to existing power plants; very expensive

IV. Indoor air pollution (IAP)

A. Introduction

   1. Indoor air can be more dangerous to human health than air outdoors

      a. More than 100 dangerous substances commonly found indoors (see Figure 5.21)

      b. Risks differ in MDCs and LDCs (see Figure 5.20)

      c. Are 10–40 times more concentrated indoors than outside

      d. IAP is #1 on EPAs list of 20 major sources of environmental cancer risk

   2. IAP more of a problem now than in past because:

      a. People spend 75%–95% of their time inside some building

      b. 1970s trend (due to the oil embargo of the Mid East) toward better energy efficiency led to houses and other buildings being build more airtight

3. The groups of people at highest risk are: very old; very young; the sick; smokers; pregnant women; have respiratory or heart problems; some factory workers

Lena Berntsen/Shutterstock.com

B. Major indoor pollutants

1. *Formaldehyde*

    a. Most common indoor toxin

    b. Found in MANY indoor products

2. *Asbestos*

    a. Not harmful unless fibers are disturbed and loosened

    b. Added to materials because of its heat resistance, strength, chemical inertness, and its ability to be woven into fabric

3. *Cigarette smoke* (first and secondhand)

    a. Contains > 4500 poisonous compounds

    b. Causes heart attacks, heart disease, strokes, cancers, emphysema

4. **Radon**—a radioactive, colorless, tasteless, odorless gas produced naturally from the break-down of uranium in rocks and soil

    a. Found everywhere; more near shale and granite deposits

    b. When seeps up through soil into the air outdoors, it dissipates and there are no ill effects

    c.   When drawn into buildings in various ways, can build up to harmful concentrations

    d.   Particles are inhaled, causing lung cancer; second only to cigarettes

    e.   Approximately 10% US homes (8 million) have radon at harmful levels

  C. Sick Building Syndrome

    1.   Refers to polluted *public* indoor environments (e.g., malls, schools, offices, etc.)

    2.   EPA estimates 20% are "sick"

    3.   Can cause a variety of symptoms, even in the same building, in different people (e.g., headaches, dizziness, forgetfulness, fatigue, etc.)

    4.   An indication a building is sick if the symptoms gone when you leave the building

  D. How to *reduce* IAP

    1.   *Asbestos*—wrap and seal it

    2.   *Radon*

        a.   Gravel layer and/or plastic sheeting under your house to prevent it from entering

        b.   Seal and caulk all openings and cracks in the foundation

        c.   Install vent pipe from basement through roof, using a fan if needed

    3.   *All types:*

        a.   Increased ventilation with fans, open windows, and doors

        b.   Various common house plants absorb some indoor pollutants

V. **Stratospheric ozone depletion**

  A. Ozone is a natural component of the stratosphere found between 10 km—50 km upward, where it is created and destroyed naturally at an equal rate

  B. Most serious threat to the ozone layer

    1.   *Chlorofluorocarbons*—CFCs family

    2.   First synthesized in 1928 to use as coolant in refrigerators

3.   1930s, General Motors developed CFCs as a propellant for their spray paint

FORMATION OF OZONE

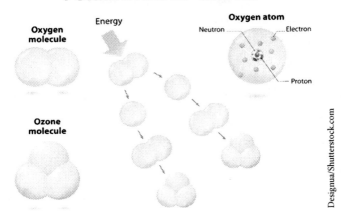

Designua/Shutterstock.com

C.   *Advantages* of CFCs

1.   Very versatile and inexpensive to produce

2.   Odorless and nontoxic

3.   Non-flammable and noncorrosive

4.   Chemically inert (stable)

a.   Not water soluble

b.   Does not combine easily with other chemicals

c.   Therefore, thought to be totally harmless to the environment and atmosphere

D.   CFCs in *stratosphere vs troposphere*

1.   F.S. Rowland and M.J. Molina at Univ. of CA—Irvine in 1974 discovered that the CFC molecule remains intact near the surface, until

2.   It diffuses slowly out of the troposphere up into the stratosphere, after months to years, where the CFCs are broken down by ultraviolet (UV) solar rays

3.   The chlorine atom released from the CFC molecule destroys an ozone molecule before being "freed" in the process to continue ozone destruction

E. Potential *atmospheric effects* of ozone depletion

1. If less UV (heat energy) is absorbed in the stratosphere because there are fewer ozone molecules to do so, the *stratosphere will become cooler.*

2. If more UV reaches the surface because less is absorbed by ozone molecules, the *troposphere will become warmer*

F. *International response* to discovery of stratospheric ozone depletion

1. 1987—50 countries signed Montreal Protocol which promised to stop the production of CFCs by 2000; was accomplished

2. Biannual meetings are held to discuss other ozone-depleting substances

3. Currently, 180 countries have signed, including China and India

G. *CFC alternative*

1. Major alternative *HCFCs*—hydrochlorofluorocarbons

2. A temporary, short-term replacement to be phased out by 2030

3. Problems: still contains chlorine; expensive to produce; flammable and toxic; breaks down so not suitable for insulation

H. What can WE do to reduce ozone depletion

1. Repair and maintain automobile air conditioners

2. Service and properly dispose of ACs and refrigerators

3. Do not use halon fire extinguishers

4. Choose home insulation without CFCs

Chapter 6

Global Climate change

I.   Introduction—Weather vs Climate

    A. Weather—short-term (daily) atmospheric conditions for a given time at a specific place

    B. Climate—a description of *aggregate* weather conditions over 30–50 years, including average temperatures and precipitation amounts as well as extremes and variability of atmospheric conditions; NOT average weather

II.   Detecting climate change

    A. Tools to monitor and collect weather and climate data directly available for only about 100 years

    B. Must depend on methods of *indirect* evidence for earlier data

        1. Comes from natural phenomena that respond to and reflect changing atmospheric conditions

        2. *Proxy data* is used to reconstruct ancient climatic conditions in an effort to understand future climate

    C. *Detection methods*

        1. *Seafloor sediments*

            a. Number and type of organisms found reflect temperature of air in past

            b. *Oxygen isotope analysis* is used ($_{18}O$ & $_{16}O$ ratio)

                (1) $_{18}O$ increased in seawater when it is *cold* and decreased when it is *warm*

                (2) $_{18}O/_{16}O$ ratio reflected in skeletal parts of micro-organisms as shells form

            c. Acquires information from up to *100 million* years ago

        2. *Ice cores* (see Figures 6.6 and 6.7)

            a. Snow that falls in polar regions becomes part of the ice pack

            b. More $_{18}O$ is evaporated from oceans when temperatures are *high*; less when it is *low*

            c. Therefore, $_{18}O$ more abundant in precipitation (snow) during warm times

            d. Acquires information up to *200 thousand* years ago

3. *Tree rings*

   a. Most accurate method

      (1) Reflects annual climatic conditions

      (2) Characteristics of the tree rings such as size and density, reflect conditions including climate, especially precipitation

   b. Use dead and living long-lived trees, such as CAs bristlecone pine

   c. Acquires information up to *10 thousand* years ago

4. *Historical records and documents*

   a. Diaries, journals, ship records, letters, art, and so on

   b. Records of crops, floods, blizzards, migration of people, and so on

   c. Problem: normal weather not likely recorded, only severe and extreme

   d. Acquires information up to *1000* years ago

III. The greenhouse effect and global warming

A. Introduction

   1. Have learned that by adding pollutants to the air, humans have altered the chemical composition of our atmosphere

ParabolStudio/Shutterstock.com

   2. Chemicals called *greenhouse gases* (GHG), some natural and some human-made, trap heat near Earth's surface

3. Process called the **natural greenhouse effect**—the physical process where shortwave, solar energy is transmitted easily through the atmosphere and most of it reaches and warms the surface; longwave, terrestrial (Earth) energy is NOT transmitted easily back into space, and most of that heat is absorbed by certain gases (GHG) in the troposphere then re-radiated back toward Earth's surface, causing an increase in surface air temperatures

4. Currently, average earth temperature is ~60°F rather than zero degrees it would be without this *natural warming*

5. However, human activities that are inputting huge amounts of GHGs are causing an *enhancement of the natural greenhouse effect,* which is leading to **global warming**—an increase in average global air temperatures

6. Past and current climate data are demonstrating that the temperature of the earth is increasing, especially since the start of the Industrial Revolution (see Figure 6.10)

B. *Major greenhouse gases* (GHGs)

1. *Natural*—water vapor

2. *Carbon dioxide* (CO_2)—50%–60% of total GHGs

   a. Major sources: burning fossil fuels; deforestation

   b. Increasing ~5% per year (see Figure 6.2)

   c. In United States, *1/3 each of total CO_2* comes from :

   (1) Transportation

   (2) Industry

   (3) Buildings (esp. heating and cooling)

3. *Methane*—12%–20% of total GHGs

   a. Major sources: livestock; rice paddies; decay in landfills; coal burning; termites

   b. Increasing ~1% per year (see Figure 6.2)

   c. Although less abundant than CO_2, is a better, *more efficient* GHG because each molecule of methane can absorb *25 times more* heat energy than CO_2 (see Table 6.1)

4. *CFCs*—15%–25% of total GHGs

    a. Major sources: foaming agents; solvents; refrigerants; spray propellant in cans

    b. Increasing ~5% per year

    c. Although less abundant than CO_2, CFCs are *super* GHGs with each molecule absorbing *15,000 times* more heat energy than CO_2 (see Table 6.1)

5. *Nitrous oxide*—5% of total GHGs

    a. Major sources: burning fossil fuel; nitrogen-based fertilizers

    b. Although less abundant than CO_2, is a *better*, more efficient GHG because each molecule of nitrous oxide can absorb *300 times* more heat energy than CO_2 (see Table 6.1)

    c. Increasing ~0.2% per year (see Figure 6.2)

C. *Potential results* of global warming (see Figure 6.9)

1. *Unequal heating across the earth* (see Figure 6.11) (US—see Figure 6.17)

    a. *Greatest* increases in average temperature expected at *high (polar) latitudes* (see Figure 6.16) and in the *middle of continents*

    b. *Least* increases in average temperatures expected at *the low (tropical) latitudes* and along *coasts*

    c. Overall, average surface temperatures on Earth expected to increase up to *1°F in 20 years* and up to *7°F by end of 21st century*

    d. Ocean temperatures also increasing, which may lead to stronger, longer-lasting hurricanes

2. *Changing precipitation patterns* (see Figure 6.12)

    a. Predicted to intensify on-going regional changes occurring over last century

        (1) The South Western United States where droughts are becoming more frequent and severe

        (2) The Midwest United States where flooding has been alternating with droughts, and so on

    b. Generally, *increased* amounts at *high* latitudes, *decreased* in *middle and low* latitudes

    c. Over time worldwide, *precipitation totals* expected to *increase*

3. *Increasing sea level*
   a. Melting continental ice (ice caps and glaciers [see Figure 6.13])
   b. Warming water expands, causing sea level rise
   c. Rose average of 6.5" in 20th Century; estimate will rise up to 20" by end of 21st century
   d. Millions of people will be displaced as the water rises

4. *Effects on organisms and ecosystems*
   a. Coral reefs dying
   b. Changes in timing of biological phenomenon (flowering)
   c. Migration disruptions
   d. Shifts of habitats (mountain species) (see Figure 6.15)

5. *Impacts on society*
   a. *Agriculture*
      (1) Overall decrease in crop production
      (2) Shortened growing seasons
      (3) Shift in which crops can be grown where
   b. *Forestry*
      (1) Greater growth in short term
      (2) Battles: drought, fire, disease, invasive species, insects
   c. *Human health*
      (1) More heat waves expected, indicating more heat-related illnesses (e.g., heat stroke)
      (2) Increase in respiratory problems (more heat=more smog)
      (3) Expansion of tropical diseases into temperate areas by insects
      (4) Disease and sanitation issues related to flooded wastewater treatment plants
   d. *Economics*
      (1) Expected to broaden gap between rich and poor
      (2) Some will benefit from the impacts of climate change, but most will not (see Table 6.2 for impacts)

D. *Responding* to climate change

1. *Mitigation*—attempts to reduce greenhouse emissions to help lessen the severity of climate change—OR—*Adaptation*—attempts to minimize the impacts of climate change by adjusting to it

2. Development of *potential solutions*

a. *Electricity* = conservation, increased efficiency, switch to cleaner energy sources, capture carbon emissions (storage)

b. *Transportation* = increased fuel and engine efficiency (see Figure 6.19), hybrid and electric vehicles, use of mass transit and ridesharing, biking or walking

c. *Agriculture* = sustainable farming, reduction of methane from manure by using it for energy, less nitric oxide from fertilizer

d. *Forestry* = preserving existing forests, reforesting cleared areas, sustainable forestry practices

e. *Waste managers*= treat wastewater, generate energy from incinerators, capture methane from landfills

3. No "magic bullet"; will require many steps by many people in many places across society to reduce emissions; many goals (see Table 6.3)

E. *Unknowns* still exist

1. *Uncertainties* about **feedbacks**—if one thing happens, it influences the occurrence of something else; **positive** = encourage or enhance global warming (e.g., water vapor); and **negative** =discourage or reduce global warming (e.g., white clouds)

2. *Accuracy* of **climate models**—computer programs that run on the world's largest and fastest computers, combine and input massive amounts of data on what is known about weather patterns, atmospheric circulation, atmosphere-ocean interactions, feedback mechanisms, and so on to simulate climate processes (see Figure 6.8); don't include everything so are still questions:

a. Water vapor (positive feedback) vs clouds (negative feedback)?

b. Ocean's ability to absorb CO_2?

c. Amount of CO_2 land and water plants will absorb through photosynthesis?

3. Input data not perfect

   a. Incorrect instruments and movement of weather stations can introduction bias or error into the database

   b. Data are samples; not always well representative of intended area (oceans, mountains)

F. "No-regrets" policy—will help prevent global warming, but also good for the environment in other ways

   1. Phase out CFCs—destroys ozone and is most potent GHG

   2. Increase *energy efficiency* everywhere in all we do (50% wasted)

   3. Switch to *renewable energy* sources; reduces CO_2 emissions and raw mineral extraction

   4. *Decrease deforestation* and increase reforestation; absorbs CO_2, makes animal habitat

   5. *Trap and burn methane* from landfills and manure lots; keeps it out of air and is an energy source

   6. Build more *water reservoirs* for global-warming-related and/or drought-related need for additional water

Chapter 7

Introduction to the Biosphere and Biogeography—The Study of The Distribution of Organisms on Earth

OUTLINE

I. Introduction
II. Food Chains, Food Webs, and Food Pyramids
III. Biogeography
IV. Biomes

I.   Introduction

    A. Theorized life began on Earth within first 1 billion years of formation

        1. First 4 billion years, life evolved in the oceans, which filtered harmful UV rays

        2. Life emerged onto land just 500,000 years ago after enough oxygen and ozone had accumulated in the atmosphere

    B. 1% of all incoming solar radiation reaching Earth's surface used for **photosynthesis**—synthesis of sugars (glucose) from atmospheric carbon dioxide and water by **autotrophs**—organisms that produces their own food from sunlight; *oxygen* released as by-product of photosynthesis

        1. Process is source of energy for all other living thing

            a. ½ of that 1% (0.5%) solar energy is used by the plant for nutrition

            b. Other ½ of that 1% solar energy stored in tissue of plant parts as energy which is passed on when either

                (1) The plant dies and decomposes (energy into soil) OR

                (2) The plant is eaten by an animal

II.   Food Chains, Food Webs, and Food Pyramids

    A. **Food chain**—the cycling of energy through the biosphere accomplished by **sequential predation**—organisms feed upon one another, with organisms at one level providing food for organisms at the next higher level (see Figure 7.7)

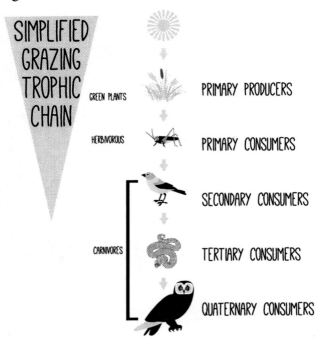

1. Each level in food chain is point of energy transfer, first from the environment (sun) to an organism (plant) then from that organism (plant) to another organism (animal); at each "link", energy transferred to the next higher organism

2. **Trophic level** (rank in feeding hierarchy) 1: *plants are producer* autotrophs that make energy available to rest of life on Earth

3. Trophic level 2: producer plants eaten by *primary consumers* called *herbivores, the plant eaters*

4. Trophic level 3: herbivores eaten by *secondary consumers* called *carnivores, the meat eaters*

5. **Omnivores**—animals that eat both plants and meat

6. Trophic level 4: carnivores eaten by other carnivores called *tertiary consumers*

7. Trophic level 5: **decomposers**—microscopic organisms that consume dead organisms and waste of living organisms, thereby returning the energy of the food chain to the soil from the waste and the dead to help grow plants

B. **Food web**—a network of who feeds on whom; an integrated food chain (see Figure 7.9)

1. Most organisms eat more than one food item and are eaten by more than organism

2. Food webs are more accurate and realistic than a simple food chain

C. **Food pyramid**—representation of transference of energy through an ecosystem based on quantity and bulk of units (see Figure 7.8)

1. Much greater numbers of lower trophic level organisms than the higher tropic level organisms they feed

2. Each organism at every level uses 90% of energy acquired from food item for own sustenance, passing on only 10% to the organism that eats it

3. Therefore, lowest levels contained highest energy per item

III. **Biogeography**

A. Hierarchical biological classification subdivides **biota**—the total complex of plant and animal life (see Figure 8.2)

1. *Highest level* = Kingdoms (3)

a. *All* animals, all plants, all micro-organisms

        b.   Contain a *multitude of different kinds* of organisms in each of the *three kingdoms*

   2.   Followed by phylum (division), class, order, family, genus

   3.   *Lowest level* = **species** (singular and plural)

        a.   Organisms that look the same and are capable of reproducing fertile offspring

        b.   Contains *only one* kind of organism in a *species,* but a *multitude of species*

B.   *Environmental* influences on the distribution of terrestrial biota

   1.   *Climatic factors*

        a.   *Moisture*

            (1)   Mass of all living things 60%–95% water

            (2)   Specific composition of organisms in an **ecosystem** (all organisms and nonliving entities that occur and interact in a particular area at the same time) are very dependent on the *amount and the timing* of precipitation

        b.   *Temperature*

            (1)   Areas of extreme cold or heat have fewer species than in moderate, temperate areas

            (2)   *Range of tolerance*—upper and lower limit and optimum temperatures

            (3)   Sometimes species' range (where found) coincide with a certain maximum or minimum **isotherm**—line connecting points of equal temperature on a map

        c.   *Light*

            (1)   Influences daily and seasonal cycles and activities of plants and animals

            (2)   Seasons differ by latitude

                (a)   Temperate regions have gradual changes in temperature and light over several weeks

                (b)   Tropical regions have abrupt changes in precipitation occurring in a few days

   2.   *Soil factors*: texture, structure; chemical composition; nutrients

   3.   *Topographic factors*: altitude; aspect; slope

4. *Disturbance*—an event that disrupts the normal functioning of an ecosystem or community

   a. Wildfire is one of most frequent

   b. Types: surface; crown or canopy; ground

   c. Can lead to **secondary succession**—a series of predictable changes as an ecosystem develops over time after a disturbance (*see* Figure 7.12) (compare with *primary succession*, Figure 7.11)

C. *Biological* influences on the distribution of terrestrial biota

1. *Evolutionary development*

   a. Controlled by the environment—how an organism evolves through time depends on its specific environment

   b. Occurs through **natural selection**—gradual genetic change in a species due to traits which are advantageous for resource acquisition and survival, being favored and passed on through subsequent generations; "survival of the fittest"

2. *Migration*–periodic movement of organisms from one place to another over various time-scales

KA Photography KEVM111/Shutterstock.com

   a. *"Seasonal"*—usually annual and due to animals following the best food source; three types—repeated return; one return; multiple generations

   b. *Environmental*—very gradual movement north or south due to drastic climate change, for example: ice age

      c. Animals are *active* migrators; plants are *passive* migrators and depend on wind, water and animals to spread seeds

  3. *Reproductive success*

      a. Each breeding group of plants and animals are in a "reproductive contest"

      b. Is competition among individuals of the same and of different species to acquire sufficient resources to breed and raise young

  4. *Displacement*—the natural, gradual removal of a species from an area due to the invasion of a more superior species, often one which is a better competitor for available resources

      a. **Plant succession**—a *type of displacement*; one type of vegetation is naturally replaced by another over time, usually due to changing environmental conditions

          (1) For example: **lake infilling**—accumulation of input sediment and debris from runoff in the bottom of lakes may eventually fill them completely over 100s–1000s years

  5. *Extinction*—the *permanent* elimination of a species of plant or animal; for example: dinosaurs 65 mya

D. *Human* influences on the distribution of terrestrial biota

  1. Affect habitats and species; are extreme, extensive, and rapid

  2. *Destruction or degradation* of habitat

  3. *Fragmentation* (into smaller pieces) of habitat

  4. *Substitution* of species

  5. *Simplification* of species—*monoculture*

  6. *Translocation* of species—introduced or invasive species (see Figure 6.13)

      a. Called *exotics* in new place

      b. Why is introducing species to new places a problem?

          (1) No predator to keep populations under control in new place

          (2) Better competitor for available resources, displacing natives

          (3) Predator that prey species are not adapted to

E. **Restoration ecology**—study of conditions of ecosystems as they existed before humans altered them by using ecological restoration

IV.   **Biomes**—10 major regional complexes of similar communities recognized mainly by its dominate plant type and vegetation structure (see Figures 7.15 and 7.16)

    A.   Temperate deciduous forest (see Figure 7.17 a and b)

    B.   Temperate grassland (see Figure 7.18 a and b)

    C.   Temperate rainforest (see Figure 7.19 a and b)

    D.   Tropical rainforest (see Figure 7.20 a and b)

    E.   Tropical dry forest (see Figure 7.21 a and b)

    F.   Savanna (see Figure 7.22 a and b)

    G.   Desert (see Figure 7.23 a and b)

    H.   Tundra (see Figure 7.24 a and b)

    I.   Boreal forest (see Figure 7.25 a and b)

    J.   Chaparral (see Figure 7.26 a and b)

Tom Reichner/Shutterstock.com

Chapter 8

Biodiversity—the variety of plants, animals, and micro-organisms on Earth and Wildlife—all free, undomesticated species of plants and animals

I.   Introduction

    A. *Types* of biodiversity

        1. Species diversity (species richness)—the *total number* of species in a given area (see Figure 8.1b)

        2. Genetic diversity—possessing *high variation in DNA* composition among individuals within a species (see Figure 8.1c)

        3. Community (ecosystem) diversity (species evenness)—the *number* of different *groups of species* in a habitat (see Figure 8.1a)

        4. Habitat diversity—the number of different *kinds of living environments* in a given area

    B. *Number* of species

        1. Have described 1.8 million species; a few thousand more discovered

        2. Estimates of actual totals range from 3 million to 100 million, with 14 million widely accepted

        3. Number of species are unevenly distributed through groups (see Figure 8.4) as well as types of habitats (e.g., tropical forest VS North Pole)

    C. *Benefits* of high biodiversity

        1. Provides food (see Figure 8.15), fuel, fiber, and shelter

        2. Purifies air and water; Detoxifies air and waste

        3. Stabilizes Earth's climate; Moderates floods, drought, and temperatures

        4. Cycles water and nutrients; renews soil fertility; pollinates plants including crops

        5. Controls pests and disease

        6. Maintains genetic resources for a variety of crops, livestock, and medicines (see Figure 8.16)

        7. Provides cultural and aesthetic values

        8. Gives us the means to adapt to change

        9. Provides ecological health

      10. Ethics—respect for other life forms

II.   **Biodiversity *loss***

    A. Extinction and **extirpation**—disappearance of a species over most of its range (see Figure 8.3)

    B. Extinction natural process, occurring since life on Earth

        1. If did not occur through natural selection and disturbances, better adapted species could not evolve

        2. Estimated 99% of all species that ever existed are extinct; remaining 1% is what is on Earth now!

    C. *Characteristics* of extinction-prone animal species

        1. Low population due to ≥1 of the following:

            a. All individuals or breeding population killed

            b. Low reproductive rate

            c. Low species density

            d. Inbreeding between relatives

        2. Very specialized diet

        3. Large body size

        4. Need for specific environment

        5. Cannot adapt to environmental or human-caused change

D. *Background extinction rate* (prior to humans)

    1. Five major extinctions in past 440 my, each eliminating at least ½ of all global species, land and water (see Figure 8.5)

    2. Human-caused extinctions have occurred since prehistory, currently at a rate 100–1000 times faster than the background rate (see Figure 8.7)

    3. Today, 75% all plant and animal extinctions attributed to human action or activity

E. Major *human factors* causing extinctions

    1. Degradation, destruction, and fragmentation (see Figure 8.10) of habitat (see Figure 8.9)

    2. **Commercial market hunting** (overhunting also)—killing of animals for profit from sale of their fur or other parts

    3. Predator and pest control

    4. Wildlife trade—medical research; small zoos; pets; "medicinal purposes" (see Figure 8.11)

    5. Introduced/invasive species

    6. Other—for example: all pollution, ozone depletion, climate change (see Figure 8.13)

F. *Legislation* protecting wildlife

    1. Early laws in the United States

        a. 1874—to protect bison, failed

        b. 1890s—hunting and fishing restrictions

        c. 1900—**The Lacey Act**—prohibits the transport of live or dead animal or their parts across state borders without a permit; still in effect today

    2. The Endangered Species Act (1973)

        a. Is the primary legislation protecting biodiversity in the United States

        b. Forbids government or private citizens from destroying endangered species and their habitats, and prohibits trade in products made from endangered species

3. *International Union for the Protection of Nature and Natural Resources*, founded in 1934

    a. Current mission—to assign "species of concern" to one of five categories:

        (1) **Endangered**—wild species with so few individual survivors that the species could soon become extinct in all or most of its range

        (2) **Threatened**—wild species still relatively abundant in its natural range but is likely to become endangered because of a decline in numbers

        (3) **Rare**—in no specific danger, but contain few individuals, making them prone to extinction; involves almost all small island species

        (4) **Out-of-danger**—those species who used to be endangered or threatened but have been rescued from risk

        (5) **Indeterminate**—species that are susceptible to being in danger, but we do not have enough information to place them in another category (e.g., how many are there?)

    b. 3000 species nominated for protection nationally; ~1100 in PA

4. **Convention on International Trade in Endangered Species of Wild Flora and Fauna (CITES)**—1973 agreement concerning worldwide protection of endangered plants and animals, and regulation of trade in live specimens, body parts, and produced derived from listed species

    a. 1975, 80 countries signed; now~120

    b. 700 species with full protection; 27,000 species with partial protection—traded only under certain circumstances and with a permit

    c. Problems with CITES

        (1) Enforcement spotty

        (2) Convicted violators only pay small fines

        (3) Falsified documents

        (4) Countries can exclude species

        (5) Trade continues in unsigned countries

G. *Prospects* of the future

    1. Accomplishments

        a. Increasing protected land and water

        b. Restoration of species' populations

        c. Improved attitude toward predators

        d. Harmful chemicals banned

        e. Advances in international protection

        f. Educational programs

    2. Global problems hindering wildlife improvement

        a. Endless search for energy sources

        b. Pollution in all four spheres

        c. Balancing the importance between economic gain and nature conservation

        d. Uncontrolled human population growth

Chapter 9

Forests, Forest Management, and Protected Areas

I.   Introduction to *Forests*

    A. Cover 31% of Earth's land surface (see Figure 9.1)

    B. Provide habitat for countless organisms (see Figure 9.4)

    C. Help maintain soil, water, and air quality

    D. Play important role in cycling of energy and chemicals through biosphere

    E. Provide humans with wood for fuel, *paper*, construction products, and so on

    F. Forest types defined by predominant tree species (see Figure 9.2)

    G. Are some of the richest ecosystems for biodiversity due to their complex structure creating a wide variety of habitats (see Figure 9.3)

Quick Shot/Shutterstock.com

II.   Forest *Loss*

    A. Most logging in the United States in pine plantations of the South and conifers in the West

        1. Are frequently a monoculture

        2. "Even-aged" stands; reduces biodiversity of all organisms (see Figure 9.12)

        3. Cut after trees at peak growth; rotation time

        4. Vulnerable to pest outbreak (see Figure 9.15)

    B. "In the time it takes to read this sentence, 5 acres of tropical forest will have been cleared." !!!!! (see EnvisionIt, p. 209)

    C. 32 million acres a year worldwide are eliminated

    D. Using forest products helped American society develop

        1. Are not harvesting forests sustainably, so depleting resource

        2. Original, never-cut, virgin *primary forests* are almost gone

    E. Most of what we see are *secondary forests*, replanted after deforesting the original forest (see Figure 9.5) which have smaller and different species of trees

III. **Forest *Management***

    A. **Sustainable resource management**—the use of strategies to manage and regulate the harvest of renewable resources in ways that do not deplete them

    B. Strategies

        1. **Maximum sustainable yield**—maximize resource extraction while keeping the resource available for the future (see Figure 9.7)

            a. Cutting trees shortly after they go through their fastest growth stage

            b. Trees cut are much smaller than if they would not have been harvested

            c. While maximizing timber production over time, it eliminates habitat for many species that require large, old trees

        2. **Ecosystem-based management**—harvesting resources in ways that minimize the impact on ecosystems and ecological processes that provide the resource

            a. Challenging to implement due to complexity of ecosystems

            b. Can mean different things to different people

        3. **Adaptive management**—the systematic testing of different management approaches to improve methods over time; fusion of science and management

    C. *Recovery* from deforestation in the United States

        1. Began managing forest resources about a century ago

        2. **National forest**—public lands (~8% of land area) in many tracts spread across all but a few states; the US Forest Service

        3. Today most timber harvesting occurs on private lands, but can extract timber from publicly held state forests (see Figure 9.9)

        4. Timber removal averages 2.1% of total per year; tree regrowth outpaces removal in National Forests 11 to 1 (see Figure 9.9)

IV. **Timber harvesting systems**

    A. **Clear-cutting**—all trees cut in an area at once, leaving only stumps (see Figures 9.10 and 9.11a)

        1. Most cost effective for timber companies; worst for environment

        2. Problems include soil erosion; changes to microclimate; destruction of entire ecosystems

    B. **Seed-tree or shelter-wood**—small numbers of large trees left standing to help reseed area and/or provide shelter for seedlings (see Figure 9.11b)

    C. **Selection**—more trees are left standing than are cut (see Figure 9.11c)— best

V. **National Forest Management Act 1976**—required all national forests to have plans for renewable resource management based on multiple use and sustainability

    A. Recent revisions include:

        1. Monitoring fish and wildlife populations

        2. Maintaining or restoring habitats

        3. Protecting important human sites

        4. Manage timber harvesting and protect "old growth"

        5. Build no new roads (Roadless rule—Clinton 2001; Bush repealed it 2005; Obama reinstated it 2008)

VI. **Federally—protected Areas** (see Figure 9.8)

    A. Includes: national parks, national wildlife refuges, wilderness(biosphere reserves); 12% world's land area

    B. History in the United States

        1. 1903—President Theodore Roosevelt declared FL's Pelican Island as first federal bird preserve and 1st unit of National Wildlife Refuge System (now the smallest wilderness area)

        2. "The nation behaves well if it treats resources as assets which it must turn over to the next generation increased and not impaired." T. Roosevelt, 1910

        3. In two terms, 1901–1909, he created 51 bird refuges in 17 states, 5 national parks and 150 national forests

C.   Why create parks and reserves?

1.   Enormous or unusual scenic features inspire people to protect them (see Figure 9 17)

2.   Offer recreation, such as hiking, camping, fishing, hunting, kayaking, photography, bird-watching, and so on

3.   Generate revenue from ecotourism

4.   Protect biodiversity

5.   Offers peace of mind, wonder, or solitude

D.   Other activities on privately-owned protected areas

1.   Hunting on ½ national refuges as wildlife management tool

2.   Can get permits that allow landowners to use a particular refuge for livestock grazing, logging, military exercise, oil and gas drilling, and so on

E.   National Parks

1.   The United States first to set aside national parks for "common people"

2.   First ones: Yellowstone (1872); Yosemite (see Figure 9.17); Sequoia; King's Canyon; Mt. Rainier; Crater Lake

3.   1916 National Park Service created to "conserve scenery, wild life, and natural historic monuments"; started with15 parks and 21 national monuments

4.   Today, more cultural and historical sites (battlefields, buildings, etc.) than for natural attributes of places; 393 units, 89 million acres

5.   Degradation of our National Parks a problem

a.   Major problem is over-crowding, especially in the most famous parks

b.   Including all parks, in 1950, 50 million visitors; in 2000, 500 million visitors (10× in 50 years)

c.   More parks requiring permits to limit numbers

d.   Some provide shuttle busses so no vehicles enter park

e.   Same problems that plague urban areas occur during peak use season (vandalisms, litter, theft, pollution, traffic jams, etc.)

f.   Others include cutting trees, collecting plants, minerals, or fossils; defacing historical structures with graffiti; introducing exotic species

James Marvin Phelps/Shutterstock.com

F. **Wilderness**

1. An area of undeveloped Federal land which is affected primarily by the forces of nature, where man is a visitor who does not remain; contains ecological, geological or other scientific or historical value; it possesses outstanding opportunities for solitude

2. From colonization through the 19th century, wilderness seen as continuing source of economic growth through exploitation of it "unlimited" natural resources

3. By 1900, with the United States an urban-industrialized nation, changed the perception of wilderness; it was now in danger of disappearing

4. First "declared" wilderness Gila National Forest, 1924

5. Wilderness Act

   a. Introduced by Wilderness Society and Senator Hubert Humphrey in 1956

   b. Signed into law by President Johnson, 1964

   c. Authorized government to set aside federally-owned land that "retains it primeval character and lacks improvements or human habitation"

6. Began with 9.2 million acres controlled by the US Forest Service (Not just forests!)

7. Today, 756 units in the United States, 109 million acres, in all 50 states with ½ in Alaska

8. 200 million + more acres could be protected by the Wilderness Act in future

9. Why protect wilderness?

    a. Are only remaining areas representing long-term balance conditions between the environment and its biota

    b. Studying these natural communities can help us manage human-altered environments

Chapter 10

Managing our Waste

I. **Categories of Solid Waste**

    A. **Municipal**—nonliquid, nonhazardous refuse from households, small businesses, and institutions

    B. **Industrial**—nonliquid, nonhazardous refuse from production of consumer foods, mining, oil extraction and refining, and agriculture

    C. **Hazardous waste**—refuse that is toxic, chemically reactive, flammable, or corrosive

II. **Components of waste management (see Figure 10.1)**

    A. Minimizing the amount of waste we generate (the *waste stream*)

    B. Recover discarded materials and finding ways to recycle them

    C. Disposing of waste safely and effectively

III. **Municipal solid waste**

    A. People in the United States create 245 million tons of municipal solid waste every year, 1500 lbs per person! (see Figure 10.2a)

    B. Approximately 4.3 lbs per person per day; an increase of 62% since 1960

    C. Two to five times more than any other MDC; 5–10 times more than any LDC

    D. Household waste categories: reusable, recyclable, compostable, toxins, 1-time use items

IV. **Solid waste reduction and management**

    A. **Resource Conservation and Recovery Act, 1976**—started national solid waste management policies

    B. Where does our solid waste go? (see Figure 10.4)

        1. Landfills, recycled or incinerated

        2. Percent varies across states

        3. Nationally, progressing from putting 80% into landfills in 1988 to the EPA goal of only 45%; recycling increased from 10% in 1988 to the goal of 35%; incineration from 10% to 20% goal

    C. **Integrated waste management**—set of management alternatives with the objective of reducing the amount of waste that must be disposed of in landfills, incinerators, and other waste management facilities that are less environmentally friendly

1. **Source reduction**—the design, manufacturing and use of products to reduce the quantity and toxicity of waste produced when products are no longer usable; object is to reduce the amount of materials that enter the waste stream

2. **Precycling**—consumers making environmentally-sound choices when purchasing products (e.g., no 1-use items)

3. *Source reduction and precycling* used together represent the greatest potential for reducing solid waste *production*

4. **Incineration**—a controlled process for burning solid waste for disposal where garbage is burnt at very high temperatures ($\geq 1650°F$) leaving ash and noncombustible to dispose of in a landfill

   a. Volume reduced by 90% and weight by 75%

   b. Problems of air pollution and disposal of toxic ash

   c. **Waste-to-energy facilities**—incinerator that uses heat from its furnace to boil water to create steam that drives electrical generation or that fuels heating systems (see Figure 10.7)

      (1) Unprocessed solid waste has 40%–75% of energy in coal; potential energy

      (2) Major problems: expensive to build and operate; some air pollution

ZouZou/Shutterstock.com

5. **Sanitary landfills**—a site at which solid waste is buried or piled into mounds for disposal, designed to prevent the waste from contaminating the environment; always be best place for some things

   a. *May not be sited* on floodplains, wetlands, earthquake zones, unstable land or near airports?

   b. Must have *liners* to prevent infiltration of liquids into soil and water

   c. Must have a system to collect **leachate**—liquids that seep through liners of a landfill and into the soil and/or water underneath

   d. Operators must *monitor groundwater* for many specified chemicals

   e. Operators must meet financial assurance criteria to ensure that *monitoring of the landfill* continues for 30 years after the closure of the site

   f. Enforcers of the *Clean Water Act* closed 1000s of local landfills in the 1980s due to their noncompliance with modern landfill regulations

   (1) 1981—18,500 landfills

   (2) 1990—6300 landfills

   (3) 2007—1750 landfills (stabilized)

   (4) More will close and few new ones built

   g. *Benefits* of modern landfill

   (1) Air pollution from burning avoided

   (2) Odor, rodents, insects, and litter minimized

   (3) Leachate controlled

   (4) Low operating costs and can handle huge amounts of waste

   (5) Methane produced for energy

   (6) When filled, can be reclaimed similarly to a mining site (see Figure 10.6)

   h. Drawbacks of landfills

   (1) Waste biodegrades very slowly in them

   (2) Increased traffic; dirty and noisy

   (3) Requires huge amounts of land space

   (4) NIMBY—reduced land values

6. **Recycling**—the collection of materials that can be broken down and reprocessed to manufacture new products—and ***reuse*** (see Figures 10.2b and 10.9; Table 10.1)

   a. *Recycling and reuse* used together represent the greatest potential for solid waste *reduction*

   b. Types of recycling

       (1) Primary—involves a final product that has characteristics the same or very close to the original product

       (2) Secondary—creates materials with characteristics that are not as close to those of the original product

   c. *Steps* of recycling (see Figure 10.8)

       (1) Collection and processing the recyclable materials

       (2) Use of the materials in making new products

       (3) Consumer purchase of these products

   d. *Economic incentives* to recycle

       (1) Tax breaks for businesses using recycled products and materials (e.g., office paper)

       (2) Grants or loans available for building recycling facilities OR to establish new markets for recycled goods

       (3) Pick up recyclables for free; charge by bag for no-recyclable trash pick-up

BestPhotoPlus/Shutterstock.com

e. *Benefits* of recycling

(1) Saves money and energy

(2) Conserves resources (less to extract)

(3) Landfills used just for necessary items and so they last longer

(4) Decrease soil, water, and air pollution

(5) Reduces litter (e.g., bottle bills)

f. *Curbside* recycling programs increasing in the United States

(1) 1050 in 1988; 9000 in 2011

(2) Recycled materials sent to a *material recovery facility*

(3) From there sent to various markets for reprocessing

g. *Obstacles* to recycling

(1) Environmental and health costs of acquiring new raw materials is not included in price of products

(2) Are more tax breaks for resource extraction industries than for recycling industries

(3) Is a lack of large, steady markets for some recycled goods

7. **Composting**—the conversion of *organic waste* (e.g., food and crop waste) into mulch or humus by encouraging, in a controlled manner, the natural biological processes of decomposition

V. ***Hazardous waste* (see Figure 10.14)**

A. In households: 1.6 million tons per year; includes: (see Figure 10.13)

1. Paint, batteries, oil, cleaning agents, lubricants, and so on

2. Synthetic organic compounds such as tires, pesticides, solvents, and so on

3. Heavy metals such as lead, mercury, copper, and so on

B. *E-waste* is growing problem (see EnvisonIt, p. 239; Figure 10.12)

1. Computers, printers, cell phones, TVs, DVD and MP3 players, fax machines, IPads, iPods, and so on

2. 3 billion electronics purchased in the United States since 1980

3. 400 million discarded electronic per year; 2/3 working

4. *Four out of five go into a landfill!*

   C. Disposal methods of hazardous waste

   1. *Designated landfills*

   2. *Surface impoundments* for liquid or dissolved waste

   a. Placed in shallow depression lined with impervious material

   b. Evaporates, leaving a residue on bottom

   c. Repeated several times then removed and transported

   3. *Deep-well injection*—wells drilled deep beneath the water table into porous rock where the waste is injected (see Figure 10.15)

VI. Global waste management problems

   A. Cost of managing it (handling, transporting, disposal, treatment)

   B. Economic loss represented by the energy, raw materials, and labor in discarded items

   C. Negative environmental impacts

   D. The solid waste problem will not go away by itself!

VII. Waste-reducing principles to live by

   A. Everything on Earth is connected.

   B. There is no "away" for the waste we produce; it has to be somewhere!

   C. 3Rs: reduce, reuse, and recycle

Chapter 11

Nonrenewable Energy Sources

I.   *Introduction* **to energy**

     A.   Global energy demands increasing

         1.   In MDCs, 20% of world's population and consume 65% of the world's energy; United States = 25% of total

         2.   In LDCs, as improve standard of living

     B.   *Energy consumption* divided evenly among buildings, industry, and transportation

     C.   Electricity supplies most global energy

     D.   Energy sources we use today (see Table 11.1)

         1.   *Nonrenewable* = fossil fuels, nuclear

         2.   *Renewable* = biomass, water, solar, wind, geothermal, ocean-tides, and waves

II.   *Nonrenewable* **energy sources**

     A.   *Fossil fuel #1—crude oil/petroleum*

DEEPWATER DRILLING   LOCATING THE OIL FIELD

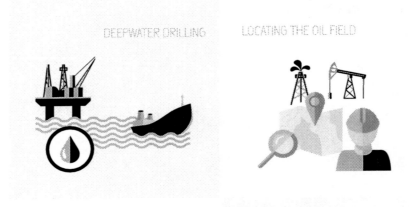

OIL TRANSPORTATION   OIL REFINERY

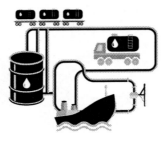

Macrovector/Shutterstock.com

         1.   Supplies 1/3 world's energy

         2.   Forms where heat and pressure transformed decomposed plant matter buried in *marine sediments* for millennia (see Figure 11.3 right)

3. *Where* oil is found (see Table 11.2)

4. Refining creates *many products* (see Figure 11.7)

    a. Highest grade, lowest boiling point = gasoline

    b. Lowest grade, highest boiling point = asphalt

5. Many products made from petroleum *we use every day* (see Figure 11.8)

6. In United States, we consume twice as much as we produce (see Table 11.5)

7. *Consumption* increased 15% in decade; 200 gallons per year per person!

8. How much oil is left? (see Figure 11.9b)

9. Most US oil in *Prudhoe Bay, Alaska*

    a. Pipeline built in 1970s to carry it to Gulf of Alaska, Valdez Terminal

    b. 800 miles long, crosses 3 mountain ranges

    c. 3 years and $8 billion to build

    d. Produced 15 billion gallons in first 30 years

10. Oil companies trying to get permission to drill in the *Alaska National Wildlife Refuge*

    a. Some believe we *should drill* because:

        (1) United States needs oil to help us become more independent of foreign oil

        (2) Price increases in recent years is economic incentive to develop domestic oil reserves

        (3) Will bring job and dollars to Alaska

        (4) New tools require fewer exploratory wells

        (5) Affected land area small compared to total

        (6) New drilling practices have less impact on the environment

    b. Some believe we *should not drill* because:

        (1) Advances in technology irrelevant; drilling would forever change the pristine environment

        (2) Intensive activity would disrupt wildlife

(3) A 1-km ice road requires 1 million gallons of water

(4) Heavy vehicles used for exploration permanently scar the ground

(5) Accidents happen

(6) Oil development damaging due to involving complex of people, vehicles, equipment, pipelines, and support facilities

11. Advantages and disadvantages to oil (see booklet)

B. *Fossil fuel #2—natural gas*

1. Supplies 1/5 of world's energy

2. Propane, butane, and methane mixture

3. Is *lighter than oil* and forms over oil deposits; collected or burned off

4. Is possible *transition fuel* from use of other fossil fuels to alternative energy sources because emits ½ CO_2 than coal and 2/3 of oil

5. US consumption about equal to its production (see Table 11.4)

6. Believe to have ~60 year supply

7. *Where* natural gas is found (see Table 11.2)

8. Relatively new source being exploited, much of it in PA, is *Marcellus Shale* extracted by hydraulic fracturing (fracking) and horizontal drilling

   a. Drill vertically 5000–8000 ft. then horizontally for up to mile

   b. Shale fractured by injecting water and sand under incredibly high pressure into the formation

   c. Makes microscopic fractures in the shale, allowing gas to escape, and the sand used to hold those fractures in place to allow the gas to continue to escape.

9. *Fracking issues*

   a. Immense volumes of wastewater returned to surface by process that contain salt, radioactive elements like radium and toxic compounds like benzene from deep underground

   b. Often sent to wastewater treatment plants not designed to handle all the contaminants and don't check for radioactivity

   c. PA very impacted as millions of gallons of drilling wastewater goes to treatment plants; "treated" water released into rivers

10. Advantages and disadvantages of natural gas (see booklet)

C. *Fossil fuel #3—coal*

   1. Supplies ¼ world's energy

   2. Most abundant, widely distributed fossil fuel

   3. Forms where heat and pressure transform decomposed plant matter *on land* for millennia (see Figure 11.3 left)

   4. US consumption about equal to production (see Table 11.3)

   5. *Where* coal is found (see Table 11.2)

   6. Used for electricity at most power plants (see Figure 11.4)

   7. *Types* of coal: peat; lignite; bituminous, anthracite (most value and rare)

   8. Estimate supply of coal will last few hundred years

   9. *Mountaintop removal mining*

      a. Used in Appalachian Mountains on immense scale

      b. Essentially, the mining company scrapes off entire mountaintops to acquire coal in small seams (up to 250'!)

      c. Most efficient and profitable for mining companies

      d. Increasing demand from China and India has doubled the coal price in central Appalachia

   10. Advantages and disadvantages to coal (see booklet)

D. *Other* fossil fuels

   1. *Oil (tar) sands*—deposits of moist sand and clay containing a thick form of petroleum which is rich in carbon and poor in hydrogen (see Figure 11.10)

      a. Are crude oil deposits that are degraded or chemically altered by water erosion and/or bacterial decomposition

      b. Removed by *strip mining*

      c. Sent to specialized refineries where it is upgraded into synthetic crude oil

      d. ¾ of it in Venezuela and Alberta, Canada

   2. *Oil shale*—sedimentary rock filled with organic matter processed to produce liquid petroleum

      a. Formed same way as crude oil but results when organic matter is not buried deep enough or is not subjected to enough heat to form oil

      b. Removed by *strip or subsurface mining*

      c. 40% global reserves in United States, mostly on federal land in CO, WY & UT

E. *Nuclear energy*

1. *History* of nuclear energy in the United States

a. Price–Anderson Act, 1957—removed all liability for the utility industry from any accident occurring in a nuclear power plant; became the US government's responsibility

b. *1st nuclear power plant* in United States build in Shippingport, PA in 1957

c. Construction accelerated in 1960s and 1970s

d. No new plants built in United States since *Three Mile Island accident, 1979*

e. In 1980s and 1990s share of electricity by nuclear power *increased* as those under construction before the accident came on line; last one 1996

f. As of 2010, 104 nuclear reactors in 34 states, producing 20% of US electricity (9 in PA)

g. Possibly a new one to be built soon?

2. Nuclear power in some countries is gaining popularity and plants are being built (see Table 11.6)

3. The *process*

   a. Fuel is *uranium*, which produces a lot of energy from a very small amount

   b. Involves neutrons bombarding uranium-235 atoms, causing nuclear *fission* splitting of an atom (see Figure 11.18)

   c. As the atoms are split, large amount of heat energy is released as well as two or more neutrons, each of which strikes another atom and a chain reaction occurs

4. Small risk for large *accidents* (see text for details)

   a. Three Mile Island, United States, 1979

   b. Chernobyl, Ukraine, Russia, 1986 (see Figure 11.21)

   c. Fukushima Daiichi, Japan, 2011 (see Figure 11.22)

5. Other than threat of nuclear accident, main obstacle to building new plants is waste disposal of all the *radioactive waste* such as spent fuel rods, some of which will still be harmful in 10,000 years!

   a. Nuclear Waste Policy Act, 1982—a waste disposal program for high-level, extremely toxic nuclear waste

   b. Specified should be deposited underground in deep, geologic waste central repository

   c. *Yucca Mt, NV* in desert 100 miles from Las Vegas chosen and $13 billion spent (see Figure 11.24)

      (1) 2010, was waiting approval of Nuclear Regulatory Commission

      (2) Obama ended its support, but is likely not totally resolved

      (3) Major issue of central location is transportation of the waste from 120+ current storage areas and operating plants, involving thousands of shipments of extremely hazardous material by rail and truck over 100s of public highways in most states

   d. Currently waste stored on-site at each plant, totaling 60,000 metric tons

   e. 161 million citizens live within 75 miles of a current site (see Figure 11.23)

w

Chapter 12

Renewable Energy Alternatives

I.   Introduction

   A.   Supplies ~8% US energy; ~18% world's (see Figures 12.1b and 12.2a)

   B.   All derived from sun's energy

   C.   *Why* should we *develop* renewable energy sources?

   1.   Regenerated by sun within time useful to humans and are inexhaustible

   2.   Cause minimal to no environmental degradation

   3.   Alleviate air pollution and GHG emissions

   4.   Helps diversify our energy sources

   5.   Will be major source of employment

   D.   Why aren't renewable energy sources able to replace fossil fuels *now*?

   1.   Lack of adequate technology and infrastructure to transfer power on required scale

   2.   Far fewer subsidies and tax breaks from government than nonrenewables; only 13% of public funds (see Figure 12.3)

II.   *Types* of renewable energy sources

   A.   **Biomass**—organic plant matter produced by photosynthesis; *bioenergy* obtained by it (indirect solar because sun's energy allows photosynthesis to occur)

   1.   Supplies 50% US renewable energy; 80% of world's (see Table 12.2 and Figure 12.12.2a)

   2.   In LDCs, 35%–90% of their energy use is from fuelwood, charcoal, and manure which is used for heating, cooking and lighting (see Figure 12.4)

   3.   Biopower generates electricity with variety of sources and techniques (see Figure 12.5)

   a.   Waste products from industry processes

   b.   Bioenergy crops especially grasses used for biofuels—liquid fuels primarily for automobiles

   c.   *Biofuels* for our vehicles

   (1)   Ethanol for gasoline-powered vehicles

   (a)   1990 Clean Air Act amendments and government subsidies lead to ethanol being added to gasoline

   (b)   Mandates 36 billion gallons be produced in the United States by 2022

   (c)   Any gasoline-powered engine can run on gas with 10% ethanol (see Figure 12.6)

Stockr/Shutterstock.com

    (2) Problems with biofuels

        (a) Growing corn for ethanol causes negative impacts on environment and requires large amounts of fossil fuels

        (b) NOT energy efficient; gain only 1.5 units of energy for every 1 unit of energy used to make it!

  d. *Biodiesel* for diesel-powered engines

    (1) Produced from vegetable oil, used cooking grease or animal fat mixed with small amount of ethanol

    (2) Can run on 100% biodiesel or 20% mix

    (3) Some busses, recycling trucks, federal fleets, and so on use the blend

    (4) Most from soybeans (The United States & Brazil), oil palm (South East Asia; see Figure 9.6), or rapeseed (Europe)

4. *Advantages* of biomass

  a. Millions of acres of land unsuitable for food production could be used to grow biomass crops

  b. Can be stored and used as needed

  c. Cheap replacement for fossil fuels

  d. Turns agricultural surplus and waste into fuel

  e.   Reduces amount of organic material put into landfills

  f.   Can be used in solid, liquid, or gaseous forms for space and water heating, producing electricity, and as vehicle fuel; very versatile!

5.   *Disadvantages* of biomass

  a.   Without good management, widespread removal of trees and other vegetation can deplete soil nutrients, cause soil erosion, water pollution and loss of wildlife

  b.   Biomass has high water content (15%–90%) so has fairly low efficiency

  c.   Collection and hauling of materials expensive due to weight

B.   *Solar* energy (direct)

1.   Supplies 0.15% US energy

2.   Has grown 30% annually worldwide for 40 years

3.   Enormous potential

  a.   10 weeks of solar energy equivalent to energy stored in all known reserves of fossil fuels on Earth!

Goodmorning3am/Shutterstock.com

  b.   Estimated recoverable energy from solar energy is 75 times the present human global energy consumption!

4.   **Passive solar energy collection**—captures sunlight directly within a structure and converts it to low temperature heat, either directly or by absorbing it into an internal structure

  a. Uses south-facing windows, overhangs, heat-absorbing construction materials, and proper vegetation around the house

  b. Used in about 1 million homes in the United States

5. **Active solar energy collection**—specially designed collectors, usually located with unobstructed southern exposure, concentrate solar energy and store it as heat for space and/or water heating (see Figure 12.9)

  a. Can be added to existing structures

  b. Used in 500,000 homes in the United States; more in China and Europe

6. Keeping solar homes *cool in summer*

  a. proper landscaping

  b. window overhangs

  c. earth tubes

  d. release heat through roof

  e. open windows

  f. build all or part underground

7. **Concentrated solar power**—an array of technologies by which solar energy is harnessed from a large area and focused onto a small area in order to generate electricity

  a. Example: numerous mirrors concentrate sunlight onto a receiver atop a tall "power tower" (see Figure 12.10) and the heat is transported by air or fluids into a steam-driven generator to create electricity

  b. International Energy Agency estimates 100 mi^2 of Nevada desert could generate enough electricity to power the entire US economy

8. **Photovoltaic (PV) cells**—devise designed to collect sunlight and convert it directly into electricity

  a. Uses thin layers of silicone that emit electrons as "sunlight" hits them

  b. Constructed in modules, enclosed in plastic or glass, which can be combined to produce systems of various size (see Figure 12.11)

  c. Initially and continually used by the space program to power satellites and space vehicles

  d. Commercially used in remote areas such as oceans buoys, lighthouses, road signs, meteorological stations, emergency telephones

      e. Recently started being used to power portable refrigerators to carry vaccines required to be cold to the nomads in African countries

      f. Now, photovoltaic materials compressed into ultra-thin sheets

        (1) cheaper to produce

        (2) can be incorporated into roofing tiles that utilize the roof of a building as a platform for its own "power plant."

      g. Increased efficiency allows conversion of $\geq$ 20% of solar energy into electricity (used to be 2%)

      h. PV technology doubled every 2 years (see Figure 12.12); The United States fifth in production

  9. *Advantages* of solar energy

      a. inexhaustible, free resource

      b. uses no fuel so creates no air pollution

      c. safe and quiet; no moving parts

      d. little maintenance; no generator required

      e. creating "green" jobs; technology well developed

  10. *Disadvantages* of solar energy

      a. need total access to sunlight which is problem in high-density areas like cities

      b. Sun doesn't shine all the time anywhere so need back-up and storage

      c. high up-front cost of equipment and installation

C. *Wind* energy (indirect solar)

  1. Windmills first used in the United States in 1600s to grind grain and pump water

  2. In 1930s and 1940s, used to generate electricity for farms beyond utility poles

  3. Now called *wind turbines*

      a. average 260' high (higher better because minimizes turbulence and maximizes wind speed) (see Figure 12.14)

      b. three fiberglass blades 260' across

      c. designed to move back and forth in response to changes in wind direction so motor faces wind at all times

      d.   begin turning when wind ≥ 10 mph and stop when wind is 50 mph

      e.   20–30 year lifespan with good maintenance

4.   Windfarms—clusters of wind turbines

      a.   1st one in New Hampshire, 1981

      b.   most in CA with ≈17,000 wind turbines

      c.   for every 10,000 megawatts wind energy turbines bring on-line, we save an equivalent to 33 million tons of coal burned each year!

      d.   today, is cheapest form of alternative energy, growing at 30% per year, 10× faster than oil

      e.   in the United States could supply 1/5 of our electric demands by 2030

5.   Wind energy locations

      a.   *Off shore* (see Figure 12.17)

          (1) wind speeds average 20% greater over water than over land with less turbulence

          (2) building costs higher but higher wind speeds equal higher profits

          (3) None in the United States, but got approval in 2010 for 130 turbines off coast of Cape Cod, Mass.; population said "no way!"

      b.   *The Great Plains* (see Figure 12.19a and b)

      c.   *mountainous regions*—ridge top or through pass

6.   Global wind power growing fast, doubling every 3 years and cost decreasing (see Figure 12.15)

7.   5 nations account for ¾ wind power output (see Figure 12.16)

8.   *Advantages* of wind power

      a.   produces no emissions, preventing tons of air pollution

      b.   can be used at all scales, from a single wind turbine to 100s

      c.   creating job opportunities (see Figure 12.18)

      d.   Very efficient: produces 23× more energy than used to produce it!

      e.   ranchers can lease land to wind companies and still use it for other purposes

9.   *Disadvantages* of wind power

      a.   wind is intermittent so need back-up (batteries)

      b.   limited areas of production (see Figure 12.19)

  c. unattractive in landscape

  d. production areas not always near population centers and the highest demand so need expanded transmission networks

  e. interferes with TVs and other electronics

  f. kills bats and birds

D. *Geothermal* (NOT from the sun)

  1. from the heat of the Earth's interior as radioactive decay of elements under high pressure deep in the interior generates heat (see Figures 12.20 and 12.21)

  2. *Types*

    a. dry steam—steam with no water droplets; best and rarest

    b. wet steam—a mixture of steam and water droplets; more common; need to remove liquid

    c. hot water—most common; need to remove salt

    d. hot dry-rock—where molten rock has penetrated the earth's crust and heats the subsurface rocks to high temperatures

N.Minton/Shutterstock.com

  3. *ground-source heat pumps* use geothermal energy (see Figure 12.22)

    a. uses the natural differences in temperature between soil and air

    b. transfers *heat* from the ground into buildings or *cools* it by transferring it into the ground

        c.   over ½ million homes use it

        d.   very efficient

            (1) heat spaces 50%–70% better

            (2) cool spaces 20%–40% better

            (3) can reduce electricity use by 25%–60%

   4.   *Advantages* of geothermal

        a.   greatly reduced emissions

        b.   using it does not reduce the amount of thermal energy produced underground

        c.   energy production ½ coal and ¼ nuclear

   5.   Disadvantages of geothermal

        a.   very limited areas of use (NOT heat pumps)

        b.   need regular groundwater recharge

        c.   due to crust shifts, may not continue to produce

E.   *Water* power—2 types

   1.   *Hydropower*

        a.   produces 16% world's energy

        b.   uses power of moving or falling water (8 lb/gal)

        c.   construction of dams originally for flood control; still are

        d.   used mostly to produce electricity, sometimes as a backup

        e.   used widely but not likely to expand (see Table 12.2)

            (1) most large rivers in the United States already dammed (98%)

            (2) people more aware of negative ecological impacts

            (3) expect share of electricity from hydropower to decrease while other renewable sources increase

            (4) the United States removing obsolete dams to restore river habitats

        f.   *Advantages* of hydropower

            (1) electricity generated cheaper than from any other source

            (2) require little maintenance and have long life times (see Figure 12.8)

(3) renewable as long as it rains regularly

(4) no pollutants emitted

g. *Disadvantages* of hydropower

Constantine Androsoff/Shutterstock.com

(1) extremely high construction costs

(2) destroys habitat, blocks migration routes

(3) displaces people and floods valuable land

(4) adds heated water to rivers

(5) natural flooding cycles disrupted so floodplains don't receive nutrients in the sediments; sediments build up behind the dam in the reservoir

(6) useless during drought

2. *Ocean energy sources*

a. *Tidal energy*

(1) produced by building dams across tidal basin outlets

(2) works with the rise and fall of the tides twice a day

(3) flow of water turns the turbine to generate electricity

(4) especially useful if very large difference between high and low tide levels

        (5) Largest in France and smaller ones in China, Russia, and Canada

    b. *Wave energy*

        (1) uses floating devises that move up and down with waves

        (2) funnels the waves from large to narrow areas

        (3) use rise and fall of waves to push air into and out of chambers

        (4) various designs developed, but none tested

  3. *Hydrogen and fuel cells*

    a. like electricity and batteries, hydrogen is an energy carrier and not a primary source

    b. electricity from renewable resources (solar or wind) could be used to produce the hydrogen

    c. fuel cells would use hydrogen to produce electrical power not only for vehicles but also electronics (see Figures 12.24 and 12.25)

    d. hydrogen gas for fuel can be produced from electrolysis (split H_2 atom from O in water)

    e. *Advantages* of hydrogen and fuel cells

        (1) hydrogen most plentiful element in universe

        (2) water and heat are only waste produced

        (3) very energy efficient

        (4) silent and nonpolluting

    f. *Disadvantages* of hydrogen and fuel cells

        (1) lack of supporting infrastructure

        (2) potential leakage of hydrogen from production and transport

III. **Today's energy choices affect future generations**

  A. Should we choose complex, centralized energy production methods, or use simpler and widely dispersed energy production, or both?

  B. Which sources of energy should be emphasized?

  C. Which uses of energy should be emphasized for maximum energy efficiency?

D. How can we rely on current energy sources and provide for developing a sustainable energy policy?

E. NO EASY ANSWERS!

IV.   Conclusions about future energy sources

A. The best short-term, intermediate-term, and long-term alternative for the world is to reduce unnecessary energy waste by improving energy efficiency.

B. Total systems of future energy alternatives in the world will probably have low to moderate net useful energy yields, AND moderate to high development costs. What is selected must be chosen carefully for maximum benefit for the lease economic and environmental costs

C. We cannot and should not depend mostly on ONE nonrenewable energy resource (oil)

D. We should reduce our dependence on large power plants and get more power from available renewable and potential resources (photovoltaic cells)